Everyday Safety in Primary and Nursery Schools

Everyday Safety in Primary and Nursery Schools is a practical, easy-to-use guide to dealing with issues of everyday safety in schools. The book provides a straightforward introduction to risk assessment, which acknowledges that things can go wrong and shows how to deal with a variety of emergencies.

The book is written with day-to-day school management in mind and will greatly assist in the business of running the school safely.

Includes material on:

- assessing risk
- developing and implementing school safety plans
- dealing with emergencies effectively
- the School Health and Safety Manual

The book includes twenty-six photocopiable safety checklists to help schools develop their risk assessment procedures and systems for dealing with incidents. There is also a list of helpful organisations and useful books and resources.

This book is an essential buy for every primary and nursery school concerned that their children are safe whilst still allowing them to develop to their full potential.

Malcolm Griffin is a former headteacher who became responsible for the management of health and safety in a London Borough. He now works as an independent consultant providing advice and training to LEAs and schools. He has worked as a consultant for the Child Accident Prevention Trust and has helped with training for the Association of Teachers and Lecturers.

Everyday Safety in Primary and Nursery Schools

Malcolm Griffin

Foreword by Carol Sherriff

RoutledgeFalmer
Taylor & Francis Group
London and New York

First published 2001 by RoutledgeFalmer
11 New Fetter Lane, London EC4P 4EE

Simultaneously published in the USA and Canada
by RoutledgeFalmer
29 West 35th Street, New York, NY 10001

RoutledgeFalmer is an imprint of the Taylor & Francis Group

Typeset in Melior by BC Typesetting, Bristol
Printed and bound in Great Britain by Bell & Bain Ltd, Glasgow

British Library Cataloguing in Publication Data
A catalogue record for this book is available from the British Library

Library of Congress Cataloging in Publication Data
Griffin, Malcolm, 1946–
Everyday safety in primary and nursery schools/Malcolm Griffin; foreword by Carol Sheriff.
p. cm.
Includes bibliographical references (p.).
1. Nursery schools–Great Britain–Safety measures. 2. Primary schools–Great Britain–Safety measures. 3. School accidents–Great Britain–Prevention. I. Title.

LB2864.5.G75 2001 00-045744
363.11′9372941–dc21

ISBN 0-415-22819-0

Contents

This book provides a largely jargon-free range of ideas to help with safety in nursery and primary schools. It is not a definitive statement of the law or a legal interpretation of any Act or Regulation. It is based on experience of good practice with the intention of helping headteachers and teachers develop systems that work in their schools with their children. The author and the publishers cannot be held responsible for the outcomes of any actions or failure to act resulting from specific interpretations of the text. In the case of any doubt you are advised to seek specialist advice

Acknowledgments

This book is based on experience gained from working with many colleague and children, both in schools and in a variety of courses in which I have been privileged to discuss and explore issues with so many dedicated colleagues. I am grateful to them for all that I have learnt from them.

My gratitude also to Linda Harrison, Headteacher and National Association of Head Teachers County Secretary for Cambridgeshire, and Susan Wilks, Technical Safety Practioner with one of the London Education Authorities, both of whom read the manuscript and offered many useful suggestions to improve it, and to Carol Sherriff, Director of the Child Accident Prevention Trust, who so kindly contributed the Foreword.

Thank you also to my wife, Margaret – mainly for her patience over many years.

Foreword

Carol Sherriff, Director of the Child Accident Prevention Trust

Primary schools and nursery schools face a difficult balance in relation to keeping their young pupils safe.

Children have a right to expect the same, if not higher, attention paid to their safety that we as adults expect in our workplaces. It is not acceptable that their future health be put at risk by accidents that are predictable and preventable.

However, children do not go to school just to work. Play and activities that help them to learn about their capabilities are an essential part of schooling. These activities must contain an element of excitement and risk if children are to be allowed to develop their full potential.

Getting the balance right requires a cyclical system of discussion, careful planning and monitoring and evaluation of injuries, accidents and near misses within the school. This book will help schools do just that by introducing a six-stage risk assessment process and by providing model policies and checklists.

To get the balance right for your children may be challenging, but it is well worth rising to that challenge. The benefits of doing so are enormous, while the costs of not doing so are potentially horrendous.

Carol Sherriff
Director
Child Accident Prevention Trust

Introduction

As a reader of this book, the chances are that you are either a Head or Deputy Head, or someone who has responsibility in some way for health, safety, welfare, pastoral care and so on within your school. If not, you are probably someone with a keen interest in those areas of school life. This book will help you with the safety side of your interest. It doesn't deal specifically with health issues or safety within curriculum areas, but it does seek to address the everyday issues in primary and nursery schools which, in my experience, most worry teachers and school governors.

These days, all sorts of establishments tend to be judged by what can be measured, but measuring care is a difficult thing to do – even if we wanted to do it. Most parents will know of the hundred and one little (and big) ways in which schools help their children by taking care of everyday matters. They will have experienced the ways in which so many members of school staffs work with great dedication to combine high standards of pupil attainment with a feeling that every child is important and every child is cared for. Although this care cannot be measured, it has become part of the way that we do things. One aspect of care which can be measured is the level of accidents occurring to pupils and, we must be careful not to forget, teachers and ancillary staff. Schools with low levels of accidents – not, I hasten to point out, low levels of accident recording and reporting – can rightly claim that such simple statistics demonstrate one tangible aspect of the 'caring school'. This book is a contribution to the process of reducing the level of accidents in schools and in so doing supports all those who inhabit them.

You will probably be familiar with some of the points made – that will be because I have tried to base the book's contents on reasonable, common-sense, good practice. Some of the points might be new to you or your school – I hope that you find them interesting and helpful. Some of the points may not be there at all! Nobody sitting down to write a book like this would imagine or pretend that they could cover every point, for every situation, in every school. Good safety management depends on the adoption of an imaginative, creative, divergent thinking approach, taking into account all that you know about children, your school and what you are trying to do to answer the question – 'Thinking of the children and adults at this school, have we got adequate arrangements in place to do what we do in a reasonably safe way?' Essentially it is the process of risk assessment set firmly within the school context.

The book is arranged in eight chapters, plus the Foreword and this Introduction. Chapter 1 explores risk assessment in relation to competence and child development, and sets out a series of straightforward principles for managing safety in primary and nursery schools. Chapter 2 deals with the items which are absolutely essential – evacuation, first aid, accident and assault reporting, training, systems for recording the building and developing effective communications. Chapter 3 is devoted to playtimes, the times when most accidents occur.

Chapter 4 addresses the structure and content of a school health and safety manual. This is supported by Chapters 5 and 6. Chapter 5 is a set of 26 photocopiable checklists which can be used either as a starting point for risk assessment or as a check to be used towards the end your assessments. The checklists are not exhaustive but should help you think widely about the factors relevant to your particular situation. Chapter 6 augments the checklists by providing some starting points on issues which go beyond the 'everyday safety' concept of the book.

At the end of the book, Chapter 7 provides a list of helpful organisations and Chapter 8 a list of useful books and resources. Throughout the book, I have tended to use the word 'incident' to cover accident, incident, near miss and so on in an attempt to minimise repeated lists of alternatives.

If you need to know the correct height for a fire extinguisher or the details of the Construction (Design and Management) Regulations, then this is not the book for you. This book is meant to help you develop your practice by taking points which relate to your school and using them. Hopefully, you will also add points from you own experience to create a safety system that works for you and your children.

Chapter 1 A 'common-sense' approach to safety with young children

Somebody once said that common sense is the obvious – once someone else has pointed it out to you. This book is meant to be one based on common sense, but equally it is hoped that it contains some advice which is not completely obvious. Having said that, I do hope that you find the contents both helpful and reasonable.

I have tried to take a straightforward approach to the issue of everyday safety and only use jargon or technical terms when I can find no sensible alternative. When jargon and technical terms are used, I try to explain them. You will also notice that the book is not dependent upon references to Acts of Parliament, Regulations, Approved Codes of Practice or Guidance (these are titles of particular sorts of documents). If you were to ask The Stationery Office for an official publication about an aspect of health and safety at work you would possibly get Regulations, Approved Codes of Practice and Guidance all bound into one. You might find it helpful to get hold of some official publications – there is a list of ones which I and other people have found helpful, at the back of the book. For the most part, my experience in working in and with schools over many years has been that there is an underlying logic to the management of health and safety and that once you understand how the logic works you can usually get things right.

So how do I perceive this underlying logic? I believe that it involves what has come to be called 'risk assessment', but I like to think of it more as being 'risk assessment in the context of education management'.

Risk assessment stage one – being clear about what is happening

The first link in the risk assessment chain is being clear about what is going on in your school – how are the people coping with the range of activities within the overall teaching and learning environment? Of course, this doesn't mean just the school site, and the word 'environment' means much more than just the physical environment. You will need to start off by thinking quite widely about this. Think for instance of the different categories of people involved, the range of activities undertaken (educational, administrative, maintenance, etc.). Think of the school at different times of the day or year – early in the morning, teaching times, lunch times, after school, evening events, weekends, holiday times and any others that you can think of. You will need to consider different locations – classrooms, boiler rooms, playgrounds, fields, shared facilities (e.g. the hall), dining facilities, vehicle routes, and of course the park, the route to the local library, and so on. If this seems like a series of long lists, don't give up – you need to 'rise above' the situation so that you can make the next important decision: the identification of significant hazards.

Risk assessment stage two – hazard identification

A hazard is usually reckoned to be something which has the potential to cause harm or loss – this could be to an individual or to the school. If this were a book about risk management, I would be suggesting that you start to think about considering how curriculum delivery could be adversely affected by a deputy head being taken out of class to deal with the after-effects of a break-in. As the book is about safety management, I will concentrate on items likely to cause harm to individuals and groups of individuals.

The phrase 'significant hazard' comes from the Management of Health and Safety at Work Regulations and self-evidently involves an element of judgement about how the people you are dealing with could be hurt. If you work with very young children, then small objects will present a choking hazard. We would, however, expect that most (but not all) older primary pupils would know not to put small 'Lego' bricks in their mouths and it would be entirely possible to leave a whole tray of beads out at a staff meeting or a governing body meeting and assume that no one would put one in their mouth!

Accidents to the children in your care can to some extent be predicted by applying what we know about child development. Toddlers are likely to grab onto things for support and so face the hazard of pulling things on top of themselves. Thinking of drowning incidents, unfortunately babies are more likely to drown in baths, toddlers in ponds, and older children in canals and lakes. Those with training in, and experience of, child development are well equipped to be aware of these stages and think ahead. Unlike those who conduct risk assessments in offices and factories, the childhood professional's perception of a hazard will relate to his or her knowledge of their particular children. An electric kettle would probably not be a problem in a staffroom, but if used in, say, a year one classroom it would present a significant hazard which would require careful management.

Another problem that we need to get out of the way at this early stage is the difference between 'predictable unpredictables' and 'unpredictable unpredictables'. The first category consists of things that, whilst not knowing *when* they might happen, we can be confident *will* happen somewhere within a realistic timescale. Fire is a 'predictable unpredictable' – we don't know which school will have a fire in any one year, but we can predict that some will. It is therefore reasonable to have fire prevention and precaution arrangements in place, as well as fire alarms and emergency evacuation procedures. On the other hand, in most parts of the United Kingdom we would all accept earthquakes as being 'unpredictable unpredictables', so there is no section in this book about earthquake plans.

There is an important point to be made here – you need to think about your local conditions. This book addresses topics which I believe are of relevance to most nursery and primary schools, but only you know about the particular local circumstances which affect you. Schools which, for instance, operate on sites adjacent to airports or chemical factories need to take into account their particular significant hazards in their risk assessments.

Risk assessment stage three – thinking about likelihood

The task is to try to make some realistic estimate about how likely it is that something will happen. There are a number of factors that you can refer to when making an estimation of likelihood, which include:

- what has happened at your school over recent years. Clearly, the better your records are, the better informed you will be.
- what has happened at similar schools over recent years. Getting hold of this information is not always easy. Some local education authorities operate 'health and safety information systems' for their schools and some of these include details of noteworthy accidents. Similarly the Health and Safety Executive publishes useful information on such matters. Items published by the relevant trade unions and professional associations, as well as the *Times Educational Supplement*, also help. Networking between schools and teachers who co-ordinate health and safety is extremely helpful.
- information gathered at training events
- the collective experience of the people at your school
- brain-storming sessions in which both staff and pupils are invited to examine critically what could go wrong.

Risk assessment stage four – thinking about severity

To think about risk as simply concerning likelihood would be to take the view that tossing a coin on a 'get it right – keep the money' basis is safer than Russian roulette. (Russian roulette is the game allegedly played by placing one bullet in one of the six chambers of a revolver, spinning the chamber and then without looking, holding the gun against the temple and pulling the trigger. Five times out of six nothing happens, but when the chamber with the bullet comes to rest in front of the firing pin, the outcome is obvious). Tossing a coin leads you to lose an average of half the time, whereas in Russian roulette you lose on average one in six times. Nevertheless, most people perceive Russian roulette as riskier than tossing a coin. Likelihood is only part of the process; it is also important to think about the severity of outcome of a possible accident – how badly would someone be hurt?

Risk assessment stage five – high or low risk?

In risk assessment, we try to make some sort of realistic assessment of how likely it is that an event will occur and how severe the outcome would be if it actually came to pass. Events which are extremely likely and which would cause serious harm or loss are 'high risk'. Events which are very unlikely and which would cause very little harm or loss are 'low risk'.

Risk assessment stage six – risk reduction

Having decided how high or low the risk is, the challenge is to reduce it. Risk reduction can be achieved by:

- making an event less likely to occur, or
- making the outcome of an event less severe.

The easiest way to make sure that an event causes no harm at all is to stop doing it. If you stop using a particular substance it can't hurt anybody. This has already happened in a number of nursery and primary situations. For instance, many particularly powerful glues no longer appear in craft activities, and copper sulphate crystal gardens have long gone from classroom window sills. There remain a number of activities which cannot just be stopped altogether. In most primary schools, most accidents appear to take place on the playground during playtimes. Stopping playtimes would solve the whole problem, but this solution would be unlikely to gain popular acceptance. We are left with making accidents less likely by, for instance, effective supervision and playground rules and with making outcomes less severe by, for instance, limiting objects that children could run into, or providing safer surfaces under and around playground equipment.

Obviously it is best for everyone if we can make accidents less and less likely to occur at all, but once you have done all you can to reduce likelihood, it is important to go on and reduce severity of outcome. I always start with likelihood first, arguing that I would prefer to have no accident at all as opposed to a minor one. On the other hand, if all that can reasonably be done to prevent an accident has been done, I would prefer that, if I do have an accident, its severity will be as low as possible.

It might help to make a sequence of the activities:

- What are we doing? (being clear about what is happening)
- What can cause harm or loss to people? (hazard identification)
- How high or low is the risk? (how likely, how severe)
- Do we need to carry on doing this activity?
- Can we do it in a way that makes an accident less likely to occur? (risk reduction)
- Can we do it in such a way that if an accident occurs, we can reduce the harm experienced? (risk reduction)
- Can we put in place any extra arrangements to keep the hazard more under control?

It may be necessary to prioritise your response in order to deal with the more hazardous things first. You will also need to allocate specific tasks to particular people. In addition you will need to continue to monitor your arrangements so that you feel confident that they are having the desired effect.

Once you have reduced the risk by making it less likely that something will happen, and if it should, that the least harm will occur, and once you have made a judgement about what you can do to control the remaining risk, you will be left with an important decision to make – is what has been done good enough to protect the people at your school?

Thinking about your people

Of course, you will have been thinking about them all the time that you have been dealing with the other questions, but now comes an important one:

'Taking into account what I know about the people I am dealing with, and the hazards that they face, have I done all that is reasonable to reduce and control risks?'

So what do you know about the people?

It is probably basically two things:
1. how many and
2. how they behave, how competent they are.

Thinking about numbers, if you can reduce the number of people exposed to a risk, then that counts as risk reduction. If a similar level of risk applies to pupils and staff and you can take the risk away from the children, then you have reduced the risk. Be careful here, though, because we are concerned with the safety of everyone on site – children *and* adults.
My experience is that in primary and nursery schools, once you have dealt with maintenance issues and certain personal safety situations, most hazards affect most people. The difference is how well they are equipped to cope with them – how competent they are.

Staff competence is assisted through training and this is dealt with later in the book. When it comes to dealing with the children's competence, it helps if they have the opportunity to be involved in the development of safe practices in the school. It is also essential (and OfSTED looks for this) that they understand how to carry out curriculum activities safely. This will mean including safety guidance within the various curriculum documents that you have in school and ensuring that children follow that guidance. Just as adults need to understand the principles of risk assessment, so do children. *The Risk Pack* published by the Child Accident Prevention Trust (details at back of book) is very helpful here. Not only does it give ideas about teaching risk assessment to children, but it also gives examples of how the concepts of risk assessment can be used as part of the delivery of particular curriculum areas.

It is also necessary for staff to understand the principles of risk assessment. With this skill they are then able to help the school in the overall process of managing the school safely by providing informed insights into a process that they understand. In addition there will be times when they need to conduct a 'situational risk assessment'. The school can do much to provide organisation-based assessments, but there will come times when an individual has to look at hazards, make an estimation about possible outcomes and make a decision about how to proceed. Examples would be making a decision about an incident which occurs during an off-site activity, or deciding what to do when suspecting that there is an intruder within the staff car parking area. Being familiar with the process will not of itself make the situation better, but it should help the individual cope more effectively.

I am frequently asked how often risk assessments need to be reviewed. If there is an incident, you need to review your assessment. This may take the form of a discussion as you complete a section on a form indicating what needs to be done to prevent recurrence. If it is a serious incident, you may find it necessary to look at what you are doing in a more structured way and make dramatic changes. If many local schools are having a particular problem, you might want to check. If the law changes or if new information becomes available, you will conduct a review. Similarly, if you have significant changes, for instance building extensions or new equipment, then a review is called for. In any event, I feel that assessments need revisiting annually. Schools can change a lot when new year groups arrive. If we establish a risk assessment system that takes into account the competency of the children, then as the children change, so might our assessment. You might also need to review more frequently in the early stages of a new project. Risk assessment is very much concerned with deciding whether or not what you are doing is adequate, so it seems entirely proper to use it frequently as the project unfolds. As things settle down, you could extend the time intervals.

The task of carrying out risk assessment in a school and then managing it safely involves the processes that I have tried to describe so far in this section. In addition, it involves a clear understanding of what we are trying to achieve in schools, which includes curriculum delivery, personal development and a clear understanding of child development.

Some factors to address

To recap, when it comes to addressing the risks associated with any activity, there are a number of factors which must be addressed. These include:

- hazard identification – people need to be aware of what could hurt them, other people or cause any sort of loss
- risk assessment – having identified the hazards, it is necessary to assess the degree of risk which is attached
- likelihood – on the basis of your experience in your school and also on the experience of others in comparable settings, how likely it is that harm will occur
- severity of outcome – if an accident did occur, how serious would the harm be
- population involved – who is really at risk. Here it is helpful to think in terms of numbers of people, but also to consider their competence
- competence – just what are the differing groups of adults and children able to do?

Once these factors have been considered, it becomes possible to decide how tightly a situation needs to be controlled. A high-risk situation involving people of relatively low competence will require significant control measures to be put in place. Alternatively, a low-risk situation with highly competent people will require a proportionately lower level of control. There is a balancing act which has to be performed between risk, competence and controls.

If it is at all possible, eliminate or reduce the risk. Then consider ways of making people more competent and of providing appropriate controls. The generally accepted principles of prevention, adapted for schools, are as follows:

1. If you can, avoid a risk altogether.
2. Carry out a risk assessment of those that you cannot avoid.
3. Address risks at source.
4. Adapt activites to what you know about individuals.
5. Think how new technologies and materials can help.
6. Implement risk prevention measures as part of overall school management.
7. Give priority to measures which can protect the whole school and everyone who uses the facilities.
8. Try to ensure that both adults and children understand what they must do.
9. Make the avoidance, prevention and reduction of risks part of the culture of your school.

Chapter 2 'Basic survival' arrangements

A colleague of mine once said that 'First aid is the last resort' and I think that I agree with him. Much of this book is concerned with trying to help you prevent incidents occurring at all, but it would be foolish to pretend that nothing could go wrong.

This section is designed to provide general advice about 'basic survival' arrangements, by which I mean:

- evacuation
- first aid
- accident/assault reporting
- training.

Everyone needs to know about these things in order to keep the school running safely. When I talk to newly qualified teachers, I always ask them if they know about how these things work in their schools. If they are uncertain about any one of them, I urge them to check next day. When people ask me what they need to tell supply staff, I give them the same list (plus necessary details of medical conditions of the children in their care) as an absolute minimum.

As well as the items which everyone needs to know about all the time, there are two other basic survival arrangements that schools have found useful:

- a record of the building
- good communications.

These are both dealt with at the end of this section. Evacuations, first aid, accident/assault reporting and training are summarised in the checklists section.

Evacuation

It is easier to think of evacuation arrangements in two categories – immediate and extended. Immediate evacuations can become extended, but not all extended evacuations have to be undertaken immediately. Some examples might help. Immediate evacuations will include fires, gas leaks, certain chemical spillages, explosions and so on. The aim in these situations is to get everyone out of harm's way as quickly as possible. Extended evacuations can result from the effects of any of the causes already mentioned, but can also result from failures of services. If, for instance, the water main is damaged, storage tanks will contain water for a period of time. There will be no drinking water straight away and, eventually, no water to flush toilets. In a relatively short space of time the school will become non-operational. Similar scenarios can be deduced for electricity failures, central heating failures and so on.

Immediate evacuations

These are usually seen in the context of potential fires and therefore are often referred to as 'fire practices', but they can have other causes. *Almost* every school I have come across has addressed this issue, but regrettably *not every* school.

In the case of an immediate evacuation, everyone needs to know what they have to do. This involves having an agreed system which is practised regularly.

In the schools that I have run, I have always expected children and staff to practice in three stages:

1. During the first few days of a new term, all classes 'walk through' their evacuation procedure at a time to suit them.
2. Once everyone has completed stage 1, an evacuation practice takes place by appointment. Children are advised that the fire alarm will sound at, say, ten minutes before break. When the alarm sounds, everyone makes their way to their assembly points to be checked by their teacher (and then remain out for playtime). Those working with the youngest children will understand that even these practices will need

careful supervision. Some small children become distressed by the sound of the alarm, but it is literally vital that we do everything possible to help them overcome this problem.

3. Within a few days there is an unannounced practice.

(This three-stage programme is repeated every term.)

This system depends on a number of arrangements being in place:

1. Teachers and others in charge of children need to have knowledge of the geography of the school. I have seen sets of fire instructions in some classrooms which detail precise routes to be taken. These only work if (a) the teacher reads and understands them and (b) there is no fire or obstruction along the route. It is much better if staff have a preferred route which has been practised plus a sensible number of alternative options. There is a problem with supply staff – if they are not given quick and effective 'induction training' on how to get themselves and the children in their care out in an emergency, then both they and the children are placed at unnecessary risk.
2. Registers need to be currently accurate and available. The first task at the beginning of each session must be marking registers and getting them safely to an agreed location (usually 'the office'). Please don't be offended by me writing this, but I'm trying to be logical and have come across unsatisfactory practices in some schools.
   There has to be a system of dealing with late arrivals, 'ins and outs' for medical appointments and so on. This can be tedious and can appear to be time consuming, but it is necessary. Similarly there needs to be a system to record the adults who are on site – teachers, visitors, school meals staff and so on. One of the many harrowing aspects of the 1999 Paddington Rail Disaster was the investigators having to establish who was on the train, let alone who was injured or missing. Accurate records will reduce the need for searches and also help to maintain an acceptable level of public confidence should a disaster occur. Checking registers is the standard school system for establishing that everyone is out of a building safely. Consideration needs to be given to how registers will be made available for use at, for instance, lunch times or during games activities. Many schools provide supervisory staff with duplicate lists, but you need to beware different versions of the same list.
   One of the by-products of the attention to school security and personal safety post-Dunblane has been signing-in systems for adults. This welcome addition to normal school procedures means that if operated consistently, a complete list is available. This of course also depends on you having a signing-out system so that you know who has left your premises. It would indeed be a tragedy for someone to be injured or die

in the act of searching for a missing person who had signed in but not out. This brings us to the next point.

3. Checking that the building is empty. Departmental stores do not, of course, keep records of who is in the building (though it is possible to count the number electronically). More usually they have procedures to check that each section of the building is empty as they evacuate. In school, as well as supervising any children in their care, nominated staff will need to check specified areas, for instance, 'the teacher in room 7 will check the library area'. Bearing in mind that one of the aims is always to reduce risk, it is helpful to keep the number of these spaces to a minimum.
4. Only allowing access to spaces which are needed for the effective delivery of the curriculum. Rooms which are not in use (boiler rooms, stock cupboards) are really best kept locked. There have been examples of fires being started in such rooms by intruders and, unfortunately, by pupils. Such rooms are obvious locations for the installation of smoke alarms, especially where there are solid doors. A smouldering fire will set off a smoke alarm, but with no alarm, the same smouldering fire could burst into flame should the door be opened by a child or adult who is not aware of what is behind the door.

Extended evacuations

These might develop from an immediate evacuation or might be put into operation if there is a non-life-threatening hazardous situation, for instance water failure.

In such cases it is worth considering an emergency relocation site (this book is for the nursery and primary range – different principles would apply at secondary level). In towns, some schools have developed reciprocal arrangements with the school in the next catchment area, others with the local high school, yet others with occupiers of local community or religious buildings. In villages, it is possible to use village halls, churches and so on. Once a relocation site is agreed, it is essential that parents know about it. It would be distressing, to say the least, to arrive at school to collect a child and find the premises empty.

Extended evacuations are helped by the provision of an emergency 'box' which will include:

- registers
- first aid kit
- duplicate emergency contact lists

- foil blankets (you might evacuate during a PE lesson)
- a sign saying something like 'We have had to leave school due to an emergency – we are safe and at . . .'. This needs to be written in the language(s) that you know to be necessary.
- a mobile telephone, which is a useful addition if you are relocating to a village hall
- anything else you consider necessary, such as medicines for people with high-risk conditions.

Some of these items could be kept in the 'box' in readiness, for instance the sign; others, for instance the registers, will be in use should the incident start off as an immediate evacuation, and some other items will need to be taken from their usual location.

It is important not to forget security. The building needs to be made as secure as possible so as not to be a target for an opportunistic thief (or even a very cunning thief who has caused the emergency).

The suggestions provided earlier are obviously easiest to apply during incidents that occur during normal teaching times, but it is worth thinking about what you would do should incidents occur at less straightforward times, for instance:

- lunch times
- break times
- after school activities and clubs arranged by the school
- after school clubs arranged by third parties
- open evenings
- concerts
- school discos.

In each case, people need to know what to do should the incident occur. On a very practical note, few schools have emergency lighting systems and so a supply of powerful battery torches in the care of staff would make it possible to evacuate with less risk should the power fail during a late afternoon or evening in winter.

First aid

Put simply, every school needs to have enough first aid cover to deal with situations likely to occur at the school. This divides into two areas: equipment and trained staff.

Determining the equipment is straightforward – you need enough of the items that your first-aiders are trained to use. In most nursery and primary locations, this usually means a first aid box situated somewhere central in the building. If your site demands it, you might need to spread locations more widely. Many schools now use portable first aid kits for off-site activities or activities on remote parts of the site (if the site is big enough to have remote parts) or for use during evacuations. I am often asked about the contents of a first aid box. This varies with size, but if you buy one from a proper supplier, the box will include a contents card to enable stock checking. The stock check is for each particular box and you will need a back-up supply somewhere else in the school. It's a bit like tea in the kitchen – a packet emptied into the caddy and at least another packet in the cupboard so that you don't run out.

Not running out is the important point. A member of the office or welfare staff needs to be nominated to check first aid boxes and bags on a regular basis. Which brings me to the second most common question 'How often is it reasonable to check?' The answer, of course, is 'It depends how often you use it.' A large-ish primary school at certain times during the year when playground accidents are high will need to check at least daily. A small nursery open for cleaning and caretaking during the holiday might go for the whole break without the box being opened. When materials are needed, they must be available. This is crucial when managing first aid bags for off-site activities.

You will need to ensure that the first aid box is correctly labelled (green label, white cross and writing) and that its location is known (an item in the staff handbook or health and safety manual plus suitable signs located around the school). The suitable signs also need to state the names of qualified first-aiders.

In 1998 the DfEE issued *Guidance on First Aid for Schools*. Their risk assessment led them to the conclusion that schools need at least one first-aider on duty and available at all times that people are on the premises. By 'first-aider' they mean someone who has successfully completed a four-day training course which is approved by the Health and Safety Executive. To have someone on duty at all times requires at least three trained people – one to be on duty, one to have a break, and one to be available for absence cover. To have someone 'available' implies that they do not have responsibilities which cannot be left while they administer first aid. Some small schools report that this is quite difficult to achieve.

Most accidents in primary schools happen at breaks and lunch times; if your first-aiders are teachers they may well be 'free' but in fact taking a well earned rest from teaching. These sorts of factors have led, over recent years, to a steady increase in the practice of appointing non-teaching staff to deal with first aid.

Your risk assessment might find that during holidays and before and after school, the level of risk goes down. In such a case you might argue for an 'appointed person' (someone trained for one day on a course approved by the Health and Safety Executive) to take charge if there is an incident.

If someone is working alone, they will still need access to first aid supplies. Remembering that 'first aid is the last resort', it may be considered prudent to limit what such people are allowed to do, so as to minimise the risk of accident at times when they are alone on the premises. For instance, always operate a rule that people using access equipment (including ladders, stepladders and similar items) must only do this when there is another adult nearby.

When dealing with lettings, you need to make first aid materials available and also to provide access to a telephone (this may be by telling users to contact the caretaker in an emergency).

Finally, a first aid logbook is essential. It need not be very complicated, but will be invaluable should there be some sort of claim at a later date. The common headings for a first aid logbook are:

1. Date of time of accident
2. Location of accident
3. Name of person in receipt of first aid (and whether they are a pupil, visitor, contractor, teacher etc.)
4. Description of injury and cause
5. Treatment given, what happened next (e.g. went back to class, taken to hospital). (Technically, you also need to record when no treatment is given.)
6. Signature of person administering first aid.

Many schools now use a simple standard form to advise parents when a child has received first aid treatment. Such a form can give an indication of the nature of the accident and injury, the time it occurred, and the treatment given, together with an invitation to the parent to seek further information from the school if they need to. Other schools send home 'head bump' letters which have a similar content plus a brief description of possible after-effect symptoms which, if occurring, should be cause for the parents to seek medical advice.

Recent changes in the law mean that someone injured as a child can make a claim for a period up to three years after they have attained their eighteenth birthday. It is advisable, therefore, to retain all first aid records (and accident records) for twenty-one years. This will mean that you will have records to respond to all claims (including any made by a person injured while visiting your school when still a baby). Some local education authorities are exploring storing records on CD-ROMs. If you work in an LEA maintained school, it is worth discussing such a move with them.

If a first-aider is not available, then it is not enough to do nothing. Doing nothing creates problems when subsequently considering concepts such as 'duty of care', 'in loco parentis', 'reasonable practicability' and so on. You may be faced with an emergency situation and have to respond by taking actions reasonable for a person of your degree of competence, experience and training. Subsequent to taking such actions it might be a court that makes the final decision as to whether this was too much or too little.

Accident reporting and recording

If your school is maintained by a Local Education Authority then there will be established accident and assault reporting arrangements. My own preference is for systems in which schools report to the Local Authority and then it reports to the Health and Safety Executive. This makes monitoring much more straightforward and also means that reports leaving the LEA all emanate from one source. (You will also need an Accident Record Book (BI 510) obtainable from The Stationery Office.)

Accident/assault forms can be requested as evidence should there be a claim for compensation at a later date. You need to find some way of storing records and copy reports for twenty-one years, as mentioned at the end of the first aid section. Inspectors of the Health and Safety Executive check through them, and if one is seen as worthy of further investigation, then it will become a key document in such an investigation. They need to be filled in very carefully by a senior member of staff or the head. The practice of injured parties completing forms to be countersigned by the head implies that the head agrees with the victim's account of what happened, and this may not always be the case.

Accident *reporting* is letting someone else know what has happened. Accident *recording* is keeping a track of what is happening at your school so that you can make better assessments of the risks to children and adults. You can then consider ways of changing what you do in order to make your school a safer place.

From time to time (say, termly) someone needs to analyse the records to try to identify the more hazardous locations, times, activities and so on together with identifying the groups of pupils or adults most at risk. Having identified these key factors, your risk assessments will be all the better.

Training

The Health and Safety at Work Act requires that staff be given necessary information, instruction, training and supervision. You will note that whilst the first three words are conceptually linked, they do not mean the same thing.

It is helpful to use some of the principles of 'training needs analysis'. An important part of training needs analysis is accepting that training has to be related to specific goals for the school and that this involves being specific about (a) what training needs to be provided and (b) who needs to be trained.

It is useful to think about what training is needed under the following headings:

1. Whole staff needs – training that needs to be given to everyone regardless of job function or position within the management system, updates on procedures.
2. Specific group needs – training that is designed with a particular group in mind, for instance midday assistants or newly qualified teachers.
3. Individual needs – some training will be needed for one or two people only, for instance caretakers, and is probably best delivered at courses organised for groups of schools (usually by the LEA).

In addition, you need to think about the specific types of training which might be needed:

1. Training about health and safety – this includes awareness of the legal situation for employers and staff, procedures for your school, and so on, and would always form part of new staff induction arrangements.
2. Special topic courses – these would include, for instance, asthma awareness, hygiene, personal safety.
3. Skills training courses – examples would be how to use tower access scaffolding systems safely, how to operate a swimming pool correctly, how to use portable electrical tools. Such courses are more concerned with ensuring that people are trained to carry out certain work activities safely.
4. Curriculum safety courses – examples would be how to teach gymnastics safely, how to set up play equipment correctly, and life-saving in pools. These courses can be seen as being similar to those described in item 3 but are specifically geared towards supporting the safe delivery of the curriculum.
5. How to learn safely – this would link with item 4 and provide advice about how to help children learn without putting themselves or others at risk.
6. Specialist safety courses – examples would be first aid training, managing hazardous substances, recognising poisonous plants.

You may feel that some of the groups overlap and you would be right. The object of thinking of it like this is to help you identify the right courses for the right people.

All training needs to be recorded in some way and it is best to maintain two separate lists. If there is an accident you might well be asked, 'Was this person trained to be doing what they were doing?' This question is best answered by keeping a record sheet for each member of staff detailing what training they have received and the date they received it. Some schools keep these records in a health and safety file whilst others merge them into the overall school system for recording staff training provision.

If someone is checking your school to ascertain how well you are addressing the safety of pupils and staff, you might well be asked a different question: 'Who in your school is trained to do this particular activity?' This question is best answered by the use of a health and safety training register. Such a register needs a page for each activity. The first aid page, for instance, could be headed 'First Aid' and indicate the names of all trained people with the date they qualified or had certification renewed.

Having two systems might sound like unnecessary bureaucracy, but it has been found to save time in the long run. However you decide to organise the paperwork, remember that you need always to record who provided the course, the date it was attended and the date of any required renewal or refresher training.

Recording the building

There is nothing worse than trying to turn off the water in an emergency when the only person who knows where the stopcock is happens to be away on a course. It's a good idea to record your building in a premises file of some sort which is then kept in a known central location, such as the school office.

The sorts of items that it would be helpful to include in a premises file are listed below:

1. Plan of the site and buildings suitably annotated to help identification of locations detailed in the rest of this list.
2. Fire arrangements
   a Location of fire extinguishers, blankets and so on. It helps to number these sequentially and indicate their locations on the plan. This helps to ensure that all are serviced regularly and that any replacements are made to the correct item.
   b Location of fire alarm call points and bells, sirens and so on. As above, numbering helps with maintenance and replacement. It also helps with ensuring that all call points are tested regularly.
   c Location of fire alarm control panel.
3. Services
   a Electricity – location(s) of electricity intake cupboard, location(s) of fuse box(es). If necessary, record which boxes serve which locations.
   b Gas – locations of main gas controls and any additional cut-off points (kitchens).
   c Water – location(s) of main stopcock(s), water tanks, stopcocks which control suites of sinks or WC cisterns and so on.
   d Oil – locations of storage tanks, control valves and so on.
   e Emergency lighting (if fitted) – location of batteries, sites of luminaires
4. Waste disposal
   a Location of general waste bin area.
   b Location of clinical waste storage.
   c Locations of inspection covers and rodding points.
   d Location of food waste storage area.

5. Asbestos
   a Details of all known locations on site and how the substance is encapsulated (sealed from access).
6. First aid
   a Location(s) of first aid box(es).
   b Location of list of qualified first-aiders.

Plus anything else significant to your school.

Good communications

As mentioned earlier, the Health and Safety at Work Act requires the provision of effective information, instruction, training and supervision. Health and Safety legislation also requires proper systems of consultation to ensure that everyone is involved in the health and safety issues that affect them. Good communications between staff will help reduce the risk of accidents and incidents occurring. People who understand systems because they were involved in their design and development are more likely to use them effectively.

If an incident or accident should occur in your school, and if it is of sufficient magnitude, then you will need to manage communications after the event to minimise long-term problems and also to maintain the good reputation of your school. An accident or incident dealt with efficiently and effectively can enhance a school's reputation, always assuming that it was not down to out and out negligence. The same incident dealt with in a confused manner will certainly give the impression that the school is careless and, by implication, that such carelessness could have contributed to the accident or incident.

Good communications as part of systems development

One of the common ways of investigating an accident or incident is to:

- examine the evidence associated with the event itself, e.g. equipment, location, people involved, what happened
- examine the relevant documentation in school, e.g. health and safety procedures, curriculum guidelines, standard lesson notes
- compare what was actually done with what the various documents say must be done.

Clearly if there are no documents, then the investigator would have no choice but to say that systems are inadequate. If the people involved had behaved as required by the documentation system, then the system would be scrutinised or external factors explored. If the people involved had not behaved as required by the documentation system, then their behaviour would be considered. In many cases there are items to be addressed in each of the three areas.

It is therefore important to have a written system – but it must be one that people are able to follow. Asking one person to take time to write procedures for everyone to follow is not usually the best way of achieving a system for everyone to use. This is especially true if the document writer has only limited experience of the work in question.

Experience is that a more effective way to go forward is to state what the school is trying to achieve in the field of safety of staff, pupils and visitors. Having done this, it is helpful to think about the range of activities at your school. You need to consider what it is that is happening in each area of activity, how people could be harmed and what it is that you are doing to make the risk of harm less. If you feel that what you are doing is adequate, then that is what needs recording in your procedures. In this way there is much less fear of mismatch or of people not doing what the book says. The book will describe what people are doing already, what is working satisfactorily.

If, on the other hand, you come to the conclusion that what you are doing is not adequate, then you need to change what you are doing. Once it has been changed to a system which you believe is adequate to protect people, then record the new system. Using this approach, the new procedures are developed through consultation and discussion, so there is less risk of mismatch or people not doing things.

Good communications as part of an ongoing preventative strategy

Preventative strategy requires:

- keeping all staff informed about issues concerned with safety, including the contents of any information or requirements from the LEA
- giving all staff regular feedback about the successes and failures (it will be mixed) of your system. This will include
  - information about accidents

  - information about near misses
  - outcomes of dry run evacuation practices
- involving staff in debate and discussion about what could go wrong and how, together, you can make this less likely (or the outcomes less severe).

Midday supervisors usually work limited hours and so need special consideration. Written bulletins as and when necessary and a basic information sheet to be kept with them as they work have both proved to be effective tools to improve communication when time is tight.

Should an incident occur good communications will help everyone cope

If an emergency or crisis situation should occur then the valuable work already done will make it easier to deal with.

It may be necessary to employ formal briefing sessions for your staff. If such briefings are necessary then:

- They need to be conducted in a fairly formal manner.
- They need to be fairly short and deal only with essentials.
- They need to be well prepared – a checklist helps, e.g. who, why, what, when, how, where; or progress, people, policy, points for action and so on.
- A clear structure helps, e.g.:
  - outline of issues
  - provision of relevant information
  - questions
  - answers
  - next actions

  (It helps to ask all those present to point out gaps in information, strategy and so on – if people are aware of any gaps, it is best for everyone that they say so at an early stage.)
- Make it clear what can be passed on to others, what will be said by a spokesperson, what must remain confidential.

When you deal with parents and children, it is important that messages and information be presented in a way which suits the receivers, but the content must be consistent and accurate.

When dealing with people outside the school, it is generally reckoned that it is best to have one spokesperson. This person might deal direct with the press/media/local community, but increasingly local councils are either requiring or urging schools to take advantage of the council's public relations service. If your school is able to be helped in this way, you will still need to nominate one person to deal with the public relations service.

Whoever the spokesperson is, they need to:

- be clear about what information is 'closed' (limited circulation) and what is 'open'
- have access to all necessary information within the school
- be actively provided with information, rather than having to seek it out
- either work through a public relations service or have access to advice from one.

Remember that the school secretary/secretaries will have to field a lot of questions and queries. To help them with this they need to be:

- kept informed about what is going on
- provided with annotated copies of all letters and circulars that have been distributed
- provided with details of whom to refer questions to
- able to assure enquirers that questions will be answered and to indicate a timescale within which they will call back or can be called back.

The general principle is that whoever deals with any sort of enquirer needs to know the boundaries beyond which they must not go. At the same time, the person acting on behalf of the school must be correctly informed so that they can provide accurate information to enquirers.

Chapter 3 Playtimes

Most primary and nursery heads and teachers agree that most accidents in school take place on the playground and therefore I have devoted a special section to this area. There are several ways in which the level of accidents can be decreased.

Setting aside the accidents which result from children simply 'falling over their own feet', playground hazards are often associated with the following:

- the condition of the playground itself
- the surrounding environment
- play equipment
- poor supervision
- children being on the playground when they shouldn't.

The playground itself

With regard to the condition of the playground itself, it is necessary to keep it maintained to a proper standard. This will involve regular maintenance checks by the caretaker and regular cleaning and clearing. The caretaker will need to check regularly for cracks, pot-holes and so on. Where these areas are considered to be hazardous, prompt action needs to be taken to effect repairs. A word of warning – if the playground is in a hazardous state, it is not sufficient simply to 'put remedial work in hand'. You will need to consider taking reasonable action to 'make safe'. Such action could include taking sections out of use, reducing the number of children using the playground, increasing supervision or not using the playground at all until the necessary work is done.

Depending on your location, you will need to establish performance standards for playground cleaning and clearing. Where there are trees, then the schedule will be more intensive in the late autumn. In locations near to bus stops or on routes to and from pubs and fast-food outlets, thoughtlessly deposited litter will need regular attention. In other locations you will need to check for discarded needles and syringes, aerosol cans and so on. Your experience will inform your risk assessment and help decide how often checks are needed.

The immediate environment of the playground

The points listed earlier in relation to the playground itself will also apply to its immediate environment. In addition it is necessary to watch out for plants and bushes which have overgrown and present a hazard instead of an amenity. It is worthwhile to check for poisonous plants that have 'arrived' (for whatever reason) in your grounds. Also look out for damaged fencing and windows that open onto paved areas.

Play equipment

Play equipment needs to be designed, installed and operated in accordance with relevant British Standards, and the manufacturer's instructions.

Equipment designed for home use or home-made equipment is not suitable for school use. Unique items such as feature walls and tunnels need to be considered very carefully before moving on from the design stage. It is really best to consult an expert before you invest large quantities of time or money. Just one unexpected accident can transform a 'feature' into a 'white elephant'.

British Standards give advice about heights, safer surfaces, gaps between equipment and so on. Always remember that this is a 'technical' area and that proper attention to specification will reduce the likelihood of accidents. Also if there is an accident and you have not worked on the basis of current good practice, then claims for negligence, or even prosecution, become more likely.

A starting point for a playground equipment checklist is provided in the checklist section.

Supervision

It is important to have a written set of playground supervision arrangements. What you put into them will depend on your children and your site, but you might like to consider the following points:

- How many adults are needed to supervise?
- Where should they be located?
- Should supervision be provided by staff located at fixed sites, or do they need to patrol?
- Does the level of supervision relate to risk assessment, e.g. do you need someone near the climbing frame, or by the gate?
- How does the 'plan' of your playground affect supervision? ('L' shapes are always difficult.)
- Are all supervisory staff briefed about what is allowed and what isn't? (acceptable and unacceptable games)
- Are the staff briefed about when to 'send a child in' and when to 'send for help'? Some schools provide supervisory staff with a supply of cards bearing the phrase 'I have been sent in by (name)' – this has had the effect of reducing pupils sending themselves in.
- Are the staff aware of signs that can indicate that a fight might be starting, e.g. chanting, bunching, sudden migration to locations on the playground?
- Are there accepted standards about the size of balls which can be used? At one of my schools we cut a circular hole in a board – if a ball drops through the hole it's OK, if not it's too big.
- Are staff clear about what to do if visitors or intruders arrive on the playground?
- Do you have a view about parents coming to chat to children through the fence at playtime?

Children being on the playground when they shouldn't

Clearly there are times when children should not be on the playground. The experience of many schools is that, by addressing the following points, unauthorised use of the playground can be reduced.

- Is it clear to parents that you only provide supervision on the playground and first aid cover for ten minutes before school and ten minutes after school?
- Is this supervision and first aid cover provided?
- Are unsupervised ball games forbidden before and after school?

- Is playground equipment use forbidden before and after school?
- Are parents reminded about all of this in writing at the beginning of each term and when new children join the school?

If all these points are answered 'yes' then you are probably left with the task of nagging about the odd child or group of children. My experience is that most parents see the logic of not letting their children arrive too early or stay too late and that the more you create an expectation that this is how things are to be done in order to protect children, the more the problem begins to diminish.

Chapter 4 The school health and safety manual

The law requires that all employers prepare a Statement of Health and Safety Policy together with a description of the organisation and arrangements that are in place to deliver that policy. This means that the governing body of foundation and voluntary aided schools will need to prepare a policy for their schools. As this policy is required by law, it should be presented to a regular meeting of the full governing body and, after it has been formally accepted by the meeting, signed and dated by the Chair of Governors. In community and voluntary controlled schools, the Local Education Authority employs the staff and therefore it must produce a policy. This policy will not only state its commitment to the health and safety of staff, pupils and persons affected by what the LEA does, but also describe its organisation and arrangements for health and safety. Even where the local authority is the employer, many aspects of organisation and arrangements will naturally be specific to the school. This means that whilst an LEA can make general arrangements for, for instance, accident reporting, those for items such as supervision of play areas will have to be made by the school itself. Some LEAs offer minimum standards for such arrangements whilst others leave it to an individual school to make arrangements based on their local risk assessment. It is reasonable for community and voluntary controlled schools to expect their LEA to provide a system which (a) establishes and advises schools of the standards expected and also (b) provides up-to-date information on issues and events to help inform school decision making. In practice, voluntary aided schools and foundation schools use the material provided by their maintaining LEA.

All schools will need to have some sort of health and safety manual which will act as a central reference point for all staff. The manual will usefully contain three sections:

Section One: Documents describing requirements imposed on the school by the Local Education Authority (or LEA requirements agreed to by the governors of foundation and voluntary aided schools).

Section Two: Health and safety information items. These could include leaflets from the Local Education Authority or the Area Health Authority or material provided by the DfEE or the Health and Safety Executive.

Section Three: 'How we do it here' – to be duplicated and distributed to staff. For foundation and voluntary aided schools the first part of this practical section needs to be produced via the governing body and to include their 'employer's' Statement of Policy emphasising their commitment to following the law in order to maintain high levels of health, safety, welfare and security for pupils and staff alike. In the case of community and voluntary controlled schools the policy would be provided by their employer, i.e. the Local Education Authority, and will be echoed by a statement from the governing body supporting the LEA policy and also adding its own commitment as a 'person having to some extent control of premises'.

The second part would be a statement describing organisation for health and safety. This simply needs to be a list of who is responsible to whom for what; for instance the early years co-ordinator is responsible for health and safety within the early years area and he or she reports to the headteacher; or health and safety requirements related to the safe teaching of science are included in the relevant curriculum documents prepared by the teacher responsible for science, reporting to the headteacher. This part would also include the allocation of responsibilities for separate blocks or split sites.

The third part will be a list of arrangements and would include the following list. Items in parentheses are issues which go beyond the scope of this book as set out in the introduction.

- accident reporting, recording, investigations
- (advice [sources])
- asbestos
- (consultation)
- contractors on the premises
- (curriculum safety)
- emergency procedures

- fire safety
- first aid
- hazardous substances
- housekeeping, including cleaning
- lettings
- maintenance – premises and equipment (including the testing of portable electrical equipment)
- (medical facilities and welfare)
- off-site activities
- (personal safety)
- playground safety
- (pupils with special needs)
- (school transport)
- (security)
- (stress)
- training
- vehicles on site
- waste disposal
- work experience

This list is developed from items recommended by the Education Services Advisory Committee (appointed by the Health and Safety Commission). The items in the list are expanded in the checklists section. Chapter 6 on 'Some wider issues' is meant to help you make a start on the items in parentheses. The notes together with the checklists should help you when it comes to writing out the specific arrangements for your school. Your documentation should also describe arrangements for risk assessments, inspections, monitoring, audit and annual review.

The list is not meant to be comprehensive (nor could it be). Your local authority might require that you have arrangements for additional areas. Alternatively you might be the only school with a particular feature on site or carrying out a particular activity. In such cases you will need to carry out a risk assessment and decide whether the arrangements that you have in place are adequate to protect the children and staff in your care. If the arrangements are not adequate, then it is clearly essential to change the arrangements and then record the revised arrangements in the health and safety documentation. Simply rewriting a section of the documentation will not necessarily have any effect on what is actually happening. Health and safety is a practical activity: the documentation is to describe what you are doing, not what you think you should be doing. When you come to look at activities which are peculiar to your school, reading through the checklists given later in this book should help you get a feel for the nature of the thinking process and help you develop your own material.

Part of this process involves risk assessment, so you will need to include within your arrangements a statement about how you carry out risk assessment in your school. There are a variety of ways of doing this and the HSE publishes the now famous 'Five Steps to Risk Assessment' (see Chapter 8 on 'Useful books and resources'). A sequence of considerations based on the points given in Chapter 1 'A "common-sense" approach to safety with young children' would be a good starting point. I always find it easier to concentrate on the risk assessment of activities, e.g. 'general teaching', and then, when necessary, to consider the use of certain locations, e.g. 'the use of the playground'. Risk assessment of just the playground can become simply a maintenance check and can ignore the interaction of the children with the environment. The question is not just 'is this location safe?' but also 'safe for who and what?'

Within your arrangements section you also need to indicate how you plan to check up that what is meant to happen does happen. It is helpful to include:

1. Active monitoring – Supervisory staff will be expected to check that those they supervise are carrying out their tasks diligently. For instance, if the caretaker has to check the climbing frame every day, who is checking that this is being done?
2. Reactive monitoring – This should describe your system for recording incidents of all sorts and analysing them in order to discern areas for further attention. For instance, most accidents seem to be happening at the top end of the playground – how can we reduce their incidence?
3. Inspections – About twice a year it is helpful for the school to self-check what it is doing. A checklist helps and those provided at the back of the book, suitably expanded and added to, would make a useful starting point. Share out the tasks – e.g. ask all staff to check their furniture one week, and carry out a visual inspection of power points and switches another week. In this way within six months you will check the school and have involved everyone in the process.
4. Audit – We can all convince ourselves that what we intend to happen is actually happening. To offset this tendency it is useful to ask a third party (perhaps from your LEA) to check your systems, procedures and practice and give you a report on what you are doing. Some schools have found it helpful for the LEA safety adviser or the school nurse to check through the accident book when they visit the school. Useful comments are reported from those who have taken advantage of help from such colleagues.

5. Review – Once a year you should review what you have achieved. Report to your governing body and plan what needs to be done during the next year.

A final point – when writing health and safety documents always use a clear straightforward style that people will understand easily. Also if people must do something, be sure to use the word 'must' (not 'should'), as this assists clarity and avoids misunderstandings.

To summarise:

School Health and Safety Manual – possible structure

Section One (if yours is an LEA maintained school)
LEA Standards for Health and Safety (as approved by the governing body if your school is a foundation or voluntary aided school)

Section Two
Health and safety information (this could be from LEAs, TUs, HSE, HSC, DfEE, OfSTED, etc.)

Section Three
The School System, containing:

1. Governing Body Health and Safety Policy (voluntary aided and foundation schools)*
   or
   LEA Health and Safety Policy and Governing Body Health and Safety Statement (voluntary controlled and community schools)*
2. Organisation for Health and Safety
3. Health and Safety Arrangements

* Points for a policy and an example of a statement follow.

The Education Services Advisory Committee document 'Safety Policies in the Education Sector' suggests points to be included in health and safety policies.

The following list is adapted from the HSC material; for the full text, please refer to the document.

- a statement of who the employer is
- a statement that the employer intends to provide staff with safe and healthy working conditions
- a statement that the employer intends not to adversely affect the health and safety of other people
- a statement that the employer will arrange proper consultations with staff about health and safety matters
- a statement that the employer will consult with individuals before they take on any specific health and safety tasks
- a description of how the employer accesses expert advice in relation to health and safety
- a statement indicating the employer's commitment to provide sufficient, relevant information and training for employees
- a statement indicating that the active involvement of all staff is needed to achieve a safe and healthy workplace
- a statement that the employer will do all that is reasonable to safeguard those who use the premises in addition to employees, e.g. pupils and students, visitors, volunteers, and contractors

GOVERNING BODY STATEMENT (COMMUNITY AND VOLUNTARY CONTROLLED SCHOOLS)

NB – This is not a Safety Policy

The Governing Body of . . . . School will, so far as is reasonably practicable, ensure that all activities under its control are carried out in accordance with the Health and Safety at Work etc. Act 1974, relevant regulations, approved codes of practice, guidance notes, the Safety Policy of the Local Education Authority and paying due regard to advice and information provided by the Authority's advisers.

The Governing Body will ensure, as far as is reasonably practicable, that the premises, all means of entering or leaving the premises available for use, any plant or substances in the premises, or provided for use there, are safe and without risk to health. In this respect, the governing body will comply with arrangements and procedures made by the Education Authority as part of its responsibilities as employer. In the case of a letting arranged by the Governing Body, it will ensure that appropriate health and safety arrangements are in place.

The Governing Body recognises that failure to comply with the Authority's policy on Health and Safety matters can result in the Authority arranging for remedial action to be taken and the costs involved being deducted from the school's budget.

The Governing Body will review this statement annually and if circumstances change. It will ensure that the school maintains, monitors and reviews its Health and Safety Policy including the necessary items of organisation and arrangements.

In order to assist in the discharge of its responsibilities, the Governing Body will receive a termly report from the Headteacher together with copies of all Health and Safety reports made to the Local Authority by the Headteacher. The Governing Body will also receive copies of reports on health and safety inspections and copies of all health and safety audit reports.

Signed: . . . . Chair of Governing Body,

on behalf of Governing Body

Dated . . . .

Chapter 5 Photocopiable checklists

List of checklists

The pages that follow are all set out in the same way:

CHECKLIST NUMBER

TOPIC

ACTIONS TO MAKE INCIDENTS LESS LIKELY AND/OR LESS SEVERE

- ■ ☐
- ■ ☐

COPING AT THE TIME OF AN INCIDENT

- ■ ☐
- ■ ☐

FOLLOW-UP

- ■ ☐
- ■ ☐

EXTRA POINTS FOR YOUR SCHOOL

NAME AND TELEPHONE NO. OF EXTERNAL CONTACT FOR HELP AND/OR ADVICE

All these pages are photocopiable. You can use them to check what you are doing already, add any extra points and then integrate the content into your school health and safety management system. They can be used as part of your risk assessment procedures, either as a starting point for discussion or as a check towards the end of the process. Please remember that they are not exhaustive lists; they are meant to help you and your colleagues think around the issues and so refine your risk assessments and systems.

A blank form is provided at the end of the checklists for you to photocopy and use for additional topics.

Photocopiable resource

CHECKLIST No. 1

ACCIDENT REPORTING, RECORDING, INVESTIGATION

See also 'Accident reporting and recording' (page 20), 'First aid' (page 17) and Checklist 11.

ACTIONS TO MAKE INCIDENTS LESS LIKELY AND/OR LESS SEVERE

- Do you have stocks of whatever forms your LEA requires you to use to report accidents? ☐
  OR (if not LEA maintained) do you have stocks of forms to report to the HSE?
- Do you have an Accident Book (BI 510)? ☐
- Do you have a first aid logbook? ☐
- Do all staff know how to report an accident? ☐

  NB: You will also need procedures for reporting assaults.
- Is there a system in place for reordering necessary documents? ☐

COPING AT THE TIME OF AN INCIDENT

Not relevant because reporting, recording and investigation happen after an incident.

Photocopiable resource

CHECKLIST No. 1 cont.

FOLLOW-UP

- Is someone investigating incidents/accidents/assaults to try to identify how specific ones could be avoided in the future? ☐
- Is someone analysing the data into information to try to spot trends? ☐
  - groups most at risk
  - riskiest locations
  - riskiest activities
  - riskiest equipment
  - riskiest times of day, week, year.
- Is someone reviewing practice in the light of individual cases and trends (as part of risk assessment review), amending the way things are done and recording the revised system in the Health and Safety Manual? ☐

EXTRA POINTS FOR YOUR SCHOOL

NAME AND TELEPHONE NO. OF EXTERNAL CONTACT FOR HELP AND/OR ADVICE

Photocopiable resource

CHECKLIST No. 2

ASBESTOS

ACTIONS TO MAKE INCIDENTS LESS LIKELY AND/OR LESS SEVERE

- Do you have a record of the locations of all known asbestos in your building? ☐
- Do you make this information available to anyone carrying out work which could disturb it? ☐
- Are staff aware of what to do if they discover or suspect that they have discovered asbestos? ☐
- Does the school know who will give them advice and help in the event of an asbestos related incident? ☐

COPING AT THE TIME OF AN INCIDENT

- If anyone suspects that they have discovered asbestos:

  1. They need to report it to the headteacher or nominated person. ☐
  2. The headteacher or nominated person needs to ensure that no one gets near to the suspected asbestos in a way which is potentially hazardous (you will probably need advice here). ☐
  3. The suspect material must be identified and appropriate action taken. ☐
  4. Information needs to be managed in a way which is helpful to people and does not create panic. ☐

Photocopiable resource

CHECKLIST No. 2 cont.

FOLLOW-UP

- Amend building records if necessary. ☐
- Advise governing body and LEA (if LEA maintained). ☐
- If procedures worked satisfactorily, continue; if not review procedures as part of risk assessment review. ☐

EXTRA POINTS FOR YOUR SCHOOL

NAME AND TELEPHONE NO. OF EXTERNAL CONTACT FOR HELP AND/OR ADVICE

Photocopiable resource

CHECKLIST No. 3

CONCERTS

ACTIONS TO MAKE INCIDENTS LESS LIKELY AND/OR LESS SEVERE

- Is the location the safest you can provide? (e.g. if you have more than one hall, are you using the one with best access?) ☐
- Have you checked whether or not you need an entertainment licence of any sort? ☐
- Are all fire exits clear and accessible? ☐
- If you do not have emergency lighting, do staff have suitable electric torches to help people out in the dark? ☐
- Is there an emergency evacuation procedure? ☐
- Have staff been briefed? ☐
- Who will inform the audience of emergency arrangements before the concert begins? ☐
- Are rostra sound and located so that they cannot slip and slide? ☐
- Are lights suitably stable and equipped with safety chains? ☐
- Are trailing leads minimised, taped down or in rubber covers? ☐
- Are combustibles kept away from potential sources of ignition? ☐
- Is a first-aider on duty? ☐
- Have you limited the size of the audience to how many can get out safely in an emergency? ☐

Photocopiable resource

CHECKLIST No. 3 cont.

- Joining chairs together with side clips, or alternatively plastic cable ties, will make them less likely to tip into a heap. (The ties are cheap and can be cut with a craft knife after the event – some schools use cord.) ☐
- Do you know how many people are present? (Tickets help here.) ☐
- Are the children registered? ☐

COPING AT THE TIME OF AN INCIDENT

- In an emergency, staff must take calm control straight away and demonstrate by their actions that the system works – this will reduce the likelihood of panic. ☐
- Provide necessary first aid. ☐
- Keep others safe. ☐

FOLLOW-UP

- Investigate, report, record, review risk assessments, modify practice and documentation as necessary. ☐

EXTRA POINTS FOR YOUR SCHOOL

NAME AND TELEPHONE NO. OF EXTERNAL CONTACT FOR HELP AND/OR ADVICE

Photocopiable resource

CHECKLIST No. 4

CONTRACTORS (LARGE PROJECTS)

ACTIONS TO MAKE INCIDENTS LESS LIKELY AND/OR LESS SEVERE

This is a complicated area and is very much governed by the requirements of the Construction (Design and Management) Regulations. Naturally you will need to think about some of the points given in Checklist No. 6 but most of the health, safety and welfare issues will need to have been made clear at a pre-contract meeting. These issues will have been discussed with the architect, the contractor, the headteacher, site foreman, etc., also with any Health and Safety Representatives, the Governing Body and the Safety Committee or Group. All agreements on procedures need to be written down. It then remains for the school to ensure that these agreed procedures are followed and monitored. So:

- Do relevant staff have copies of the agreed procedures as they apply to them? (Relevant staff will include all those consulted plus the deputy head.) ☐
- Have other staff been given information about the agreed procedures and how they are to work differently during the period of the contract? (these staff will include class teachers, office staff, midday assistants) ☐
- Do all staff know who they must refer to if an agreed procedure is breached or if they suspect that it is breached or if they feel things are not really working properly? (You will find that most contractors prefer to deal with one or two named people.) ☐

COPING AT THE TIME OF AN INCIDENT

- Incidents could be many and varied – you will need to have agreed emergency procedures with the contractors. ☐

Photocopiable resource

CHECKLIST No. 4 cont.

- In most cases, the contractor will be able to make equipment and materials safe. ☐
- First aid. ☐
- Keep others safe. ☐
- Evacuation procedures. ☐

FOLLOW-UP

- Work should not resume until all parties are satisfied that such a resumption is acceptable. ☐
- This will involve investigation, reporting, recording and review of risk assessment. ☐

EXTRA POINTS FOR YOUR SCHOOL

NAME AND TELEPHONE NO. OF EXTERNAL CONTACT FOR HELP AND/OR ADVICE

Photocopiable resource

CHECKLIST No. 5

CONTRACTORS (SCHOOL PHOTOGRAPHER)

It is generally the case that a school photographer is expected simply to be responsible for taking photographs. Any essential supervision, scheduling, etc. is normally reckoned to be the work of the teaching staff.

This section deals only with safety aspects of the activity.

This list is smaller than and different to the list for small maintenance jobs (Checklist No. 6).

ACTIONS TO MAKE INCIDENTS LESS LIKELY AND/OR LESS SEVERE

- Does the photographer know that she or he must sign in and out? ☐
- Is it clear that the photographer must wear an identity badge at all times? ☐
- Does the photographer know:
  - where to park (so as not to obstruct emergency routes)? ☐
  - where on the site they will need to be accompanied? ☐
  - what school facilities they can and cannot use? ☐
  - rules about smoking? ☐
  - rules about the use of radios? ☐
  - requirement that equipment and materials must never be left unattended? ☐
  - location of first-aider and first aid box? ☐
  - what to do if there is an emergency, e.g. fire evacuation? ☐
  - requirement not to be over-familiar with pupils? ☐
- Are there adequate arrangements for bringing younger siblings from other classes? ☐

Photocopiable resource

CHECKLIST No. 5 cont.

- Are there adequate arrangements for the safety of parents and pre-school siblings? ☐
- Is it clear that any arrangements for groups of children and staff to be photographed (e.g. standing on rostra) are to be agreed with the school beforehand and that pupils and staff must not be expected to balance on improvised structures, e.g. chairs stood on tables? ☐
- Are all trailing leads used by the photographer in rubber covers or suitably taped down, preferably well away from routes to be used by children and staff? ☐

COPING AT THE TIME OF AN INCIDENT

- If there is an incident of any sort, work must cease, and injured parties dealt with whilst others are kept safe. ☐

FOLLOW-UP

- Photography must not resume until the school is satisfied that this is acceptable. ☐
- Procedures must be reviewed before next session. ☐
- Both of these require investigation, reporting, recording and review of risk assessment. ☐

EXTRA POINTS FOR YOUR SCHOOL

NAME AND TELEPHONE NO. OF EXTERNAL CONTACT FOR HELP AND/OR ADVICE

Photocopiable resource

CHECKLIST No. 6

CONTRACTORS (SMALL MAINTENANCE JOBS, GARDENERS, IT SUPPORT STAFF, ETC.)

ACTIONS TO MAKE INCIDENTS LESS LIKELY AND/OR LESS SEVERE

- Is there a sign outside your school which says that all visitors and contractors must report to the school office? ☐
- Is it clear that visiting contractors must sign in and out? ☐
- Is it clear that contractors must wear identity badges at all times? ☐
- Do you have an information sheet for contractors stating how you expect them to behave on your premises? (Such a leaflet is best written in a friendly positive style, e.g. 'to help us all care for one another, please would you be sure to observe the following points . . .'). The series of points you could cover includes:
  - where to and where not to park their vehicles so as not to obstruct emergency routes ☐
  - areas where they will need to be accompanied, e.g. certain toilets and, if you have them, changing rooms ☐
  - rules about smoking ☐
  - rules about the use of radios ☐
  - requirement that tools, equipment and materials are never left unattended in areas used by children ☐
  - location of first-aider and first aid box ☐
  - what to do if there is an emergency, e.g. fire evacuation ☐
  - requirement not to be over-familiar with pupils ☐
  - requirement to sign out ☐
  - what school facilities can and cannot be used. ☐

Photocopiable resource

CHECKLIST No. 6 cont.

- If working during school hours, have you made it clear when there are likely to be large groups of children on the move, e.g. playtime? ☐
- If working during school hours, have you made it clear where pupil traffic routes could affect working arrangements, e.g. watch out for painters working over doors which are exits to the playground? ☐
- Do you have a system to inform staff when and where any work is in progress? ☐

COPING AT THE TIME OF AN INCIDENT

- Injured parties must be dealt with. ☐
- All others must be kept safe. ☐
- Any equipment or materials presenting unacceptable risks must be attended to. ☐

FOLLOW-UP

- Investigate, report, record, review risk assessment. ☐
- If school is at fault, modify arrangements, record. ☐
- If contractor is at risk, either obtain acceptable assurances that recurrence is unlikely or change contractor. ☐

EXTRA POINTS FOR YOUR SCHOOL

NAME AND TELEPHONE NO. OF EXTERNAL CONTACT FOR HELP AND/OR ADVICE

Photocopiable resource

CHECKLIST No. 7

CONTRACTORS (VISITING DRAMA GROUPS) (ALSO APPLIES TO ROAD SHOWS)

This list is smaller than and different to Checklist No. 6 for small maintenance jobs.

ACTIONS TO MAKE INCIDENTS LESS LIKELY AND/OR LESS SEVERE

- Is it clear that the members of the group must sign in and sign out? ☐
- Is it clear that, unless it is necessary to their costume, they must wear identity badges at all times? ☐
- Do they know:
  - where to and where not to park their vehicles so as not to obstruct emergency routes? ☐
  - areas where they will need to be accompanied, e.g. certain toilets and, if you have them, changing rooms? ☐
  - rules about smoking? ☐
  - rules about the use of radios? ☐
  - requirement that equipment and materials are never left unattended in areas used by children? ☐
  - location of first-aider and first aid box? ☐
  - what to do if there is an emergency, e.g. fire evacuation? ☐
  - requirement not to be over-familiar with pupils? ☐
  - what school facilities they can and cannot use? ☐
- Is all their equipment (lights, rostra, scenery) located so as not to obstruct emergency routes? ☐
- Are all their leads, etc. taped down and preferably routed away from routes to be used by children and teachers? ☐

Photocopiable resource

CHECKLIST No. 7 cont.

- Is any portable lighting equipment secure (e.g. stands weighted/ designed to prevent toppling over, spotlights fixed with safety chains)? ☐
- Do they need a changing room and do staff and children know that this is 'out of bounds' to them? ☐

COPING AT THE TIME OF AN INCIDENT

- Injured parties must be dealt with. ☐
- All others must be kept safe. ☐
- Any equipment or materials presenting unacceptable risks must be attended to. ☐

FOLLOW-UP

- Investigate, report, record, review risk assessment. ☐
- If school is at fault, modify arrangements, record. ☐
- If drama group is at risk, either obtain acceptable assurances that recurrence is unlikely or change drama group. ☐

EXTRA POINTS FOR YOUR SCHOOL

NAME AND TELEPHONE NO. OF EXTERNAL CONTACT FOR HELP AND/OR ADVICE

Photocopiable resource

CHECKLIST No. 8

EMERGENCY EVACUATION (IMMEDIATE)

ACTIONS TO MAKE INCIDENTS LESS LIKELY AND/OR LESS SEVERE

See also 'Immediate evacuations' (page 14).

- Do you have systems for all times of day? ☐
- Does everyone know what they have to do? ☐
- Have they practised it? ☐
- Are evacuation routes properly signed? ☐
- Do all staff know the site well enough to take an alternative route if necessary? ☐
- Have you got a system for dealing with temporary staff? ☐
- Have you got a system for dealing with visitors? ☐
- Are registers of adults and children on site kept up to date during the day? ☐
- Are necessary lists in the possession of midday assistants, staff teaching games activities, etc.? ☐
- How do you check that all rooms are empty? ☐
- Are rooms which are out of use kept locked? ☐
- Have you got a portable first aid kit? ☐

Photocopiable resource

CHECKLIST No. 8 cont.

COPING AT THE TIME OF AN INCIDENT

Not relevant – emergency evacuation itself constitutes the school coping with a serious incident.

FOLLOW-UP

Investigate, report, record, review risk assessment. ☐

- Discuss what happened with staff and children. Identify what worked well and what worked less well. ☐
- Invite suggestions for improvement. ☐
- Dry run modified arrangements. ☐
- Agree new way of doing things. ☐
- Record new procedure in Health and Safety Manual. ☐

EXTRA POINTS FOR YOUR SCHOOL

NAME AND TELEPHONE NO. OF EXTERNAL CONTACT FOR HELP AND/OR ADVICE

Photocopiable resource

CHECKLIST No. 9

EMERGENCY EVACUATION (EXTENDED)

See also 'Immediate evacuations' (page 14), 'Extended evacuations' (page 16) and Checklist No. 8.

ACTIONS TO MAKE INCIDENTS LESS LIKELY AND/OR LESS SEVERE

- Do you have an emergency relocation site? ☐
- Do parents, children and staff know about it? ☐
- Do you have an 'emergency box'? Contents to include:
  - registers ☐
  - first aid kit ☐
  - contact lists ☐
  - foil blankets ☐
  - suitable signs (in relevant languages) ☐
  - mobile phone ☐
- Will someone make the building as secure as possible? ☐

COPING AT THE TIME OF AN INCIDENT

Not relevant – emergency evacuation already constitutes the school coping with a serious situation.

Photocopiable resource

CHECKLIST No. 9 cont.

FOLLOW-UP

- Investigate, report, record, review risk assessment. ☐
- Discuss what happened with staff and children. Identify what worked well and what worked less well. ☐
- Invite suggestions for improvement. ☐
- Dry run modified arrangements. ☐
- Agree new ways of doing things. ☐
- Record new procedure in Health and Safety Manual. ☐

EXTRA POINTS FOR YOUR SCHOOL

NAME AND TELEPHONE NO. OF EXTERNAL CONTACT FOR HELP AND/OR ADVICE

Photocopiable resource

CHECKLIST No. 10

FIRE SAFETY

There are things that we can all do to prevent, and cope with, fire. Other things require considerable technical knowledge and expertise. The list provided must be regarded as the most basic.

If you are an LEA maintained school, your LEA may be able to provide advice related to your building. If not, please seek competent advice. In all cases the local Fire Service will help.

ACTIONS TO MAKE INCIDENTS LESS LIKELY AND/OR LESS SEVERE

Fires starting and/or spreading, check for

- Sources of ignition, e.g. heaters, equipment (normal use/careless use/accidental failure), cigarettes, matches, potential arsonists ☐
- Combustible materials, e.g.
  - loose paper in rooms ☐
  - unemptied waste-paper bins left overnight ☐
  - aerosols ☐
  - outside rubbish bins (need lids, need to be at least 10 m from school if possible and chained down). ☐
- Building features
  - areas where fire could spread easily ☐
  - areas where smoke would accumulate ☐
  - areas which would burn easily ☐
  - non-automatic fire doors fixed open. ☐

Photocopiable resource

CHECKLIST No. 10 cont.

Detection

- Check for locations in the school where a fire could start and not be noticed. Can you make things better with any of the following?
  - vision panels ☐
  - automatic detection, e.g. smoke alarms ☐
  - lock stock cupboards, etc. to deny easy access to vandals. ☐

Warning

- Check fire alarms every week, from different call points, checking all call points within thirteen weeks. If you have more than thirteen points, then some weeks you will need to check more than one. ☐
- Check that alarms can be heard in all parts of school. ☐
- Remember adults/children with hearing difficulties. ☐
- Are instructions for use posted by all call points? ☐

Means of escape

- Do people know how to get out safely? ☐
- Are straightforward instructions posted in all rooms? ☐
- Have people practised? ☐
- Are escape routes kept clear? ☐
- Do escape routes lead to a safe location? ☐

Photocopiable resource

CHECKLIST No. 10 cont.

- Do you have emergency lighting? ☐
- Are routes properly indicated with signs? ☐
- Can people get out in time? ☐

Fire-fighting equipment

- Do you have enough? ☐
- Is it properly maintained? ☐
- Do you have the correct extinguishers for the likely fire(s)? ☐
- Are they properly located? ☐

First aid

- Do you have a first aid bag to take out of the building? ☐

COPING AT THE TIME OF AN INCIDENT

Key points if a fire starts are:

- raising the alarm, calling the Fire Service ☐
- getting out safely ☐
- helping children and visitors out safely ☐

Photocopiable resource

CHECKLIST No. 10 cont.

- assembling in the safe location ☐
- using equipment correctly (if necessary) ☐
- being able to administer necessary first aid. ☐

FOLLOW-UP

After a time there needs to be a reappraisal of practice related to investigation findings. This will probably relate to the causes of the fire and how it was dealt with. Once this reappraisal is complete, physical arrangements and working practices may need to be changed along with procedures. ☐

This will involve school investigation, reporting, recording and review of risk assessment. ☐

EXTRA POINTS FOR YOUR SCHOOL

NAME AND TELEPHONE NO. OF EXTERNAL CONTACT FOR HELP AND/OR ADVICE

Photocopiable resource

CHECKLIST No. 11

FIRST AID

ACTIONS TO MAKE INCIDENTS LESS LIKELY AND/OR LESS SEVERE

- Do you have enough boxes? ☐
- Do you need portable kits? (Large site? Off-site work? Evacuations?) (See also Checklist No. 22.) ☐
- Do you carry back-up stocks? ☐
- Does someone check the box(es) and kit(s) regularly? ☐
- Are all kit(s)/box(es) clearly labelled? ☐
- Is their location signed? ☐
- Do you have enough first-aiders? ☐
- Are their names clearly displayed? ☐
- Do you have a first aid logbook? ☐

COPING AT THE TIME OF AN INCIDENT

- First aid could be needed to cope with almost all incidents. ☐

Photocopiable resource

CHECKLIST No. 11 cont.

FOLLOW-UP

- After incidents, discuss and decide whether your equipment and trained staff provide adequate cover. ☐
- From time to time, review provision as part of your risk assessment review. ☐
- Notes of investigations, reports and records will help you. ☐

EXTRA POINTS FOR YOUR SCHOOL

NAME AND TELEPHONE NO. OF EXTERNAL CONTACT FOR HELP AND/OR ADVICE

Photocopiable resource

CHECKLIST No. 12

HAZARDOUS SUBSTANCES

If you are an LEA maintained school, talk to your Health and Safety Adviser about the Control of Substances Hazardous to Health Regulations. If not, be sure to seek competent advice.

ACTIONS TO MAKE INCIDENTS LESS LIKELY AND/OR LESS SEVERE

- Have you reduced your use of hazardous substances to an absolute minimum? (Not just in classrooms; think of offices, cleaning, etc.) ☐
- Do you always deal with reputable suppliers? ☐
- Is everyone clear that manufacturer's instructions must always be followed? ☐
- Is everyone clear that substances must be used only for the purpose for which they were purchased? ☐
- Are hazardsous substances kept in suitable labelled (relevant languages and/or pictograms) containers and in a secure place when not in use? ☐
- Is everyone clear that they may bring substances onto the premises only with the permission of a nominated senior member of staff? ☐
- Do you liaise with cleaners, builders, maintenance staff, etc. about any substances they intend to use (and about any substances you use)? ☐
- Do you keep up-to-date product information/data sheets about the hazardous substances used in your school? ☐
- If any hazardous substances are to be used in a lesson, are safety arrangements included in the lesson notes? ☐

Photocopiable resource

CHECKLIST No. 12 cont.

COPING AT THE TIME OF AN INCIDENT

- If substances are left out, do not leave them or the children unattended. Send for another adult to take them away to a safe location. ☐
- If there is a spillage, keep people well away and follow the manufacturer's advice (taken from the information/product data sheets). This might include calling the emergency services. ☐
- If anyone is hurt, follow the manufacturer's advice (taken from the information/product data sheets) and seek help as appropriate. This might include calling the emergency services. ☐

FOLLOW-UP

- Record the incident in your school incident file, and investigate what happened. ☐
- If necessary, report to your LEA Health and Safety Section or the HSE depending upon your LEA's procedures. ☐
- Discuss the incident with everyone involved and come to a decision about how you could make it less likely to occur again. (Review risk assessment.) ☐

EXTRA POINTS FOR YOUR SCHOOL

NAME AND TELEPHONE NO. OF EXTERNAL CONTACT FOR HELP AND/OR ADVICE

Photocopiable resource

CHECKLIST No. 13

HOUSEKEEPING

ACTIONS TO MAKE INCIDENTS LESS LIKELY AND/OR LESS SEVERE

- Is the school as tidy as possible? ☐
- Is lighting adequate (especially on staircases)? ☐
- Is the school clean? ☐
- Is rubbish disposed of safely? ☐
- Are furniture and equipment kept tidily? ☐
- Are floors free from holes, damage, trip hazards, etc.? ☐
- Are spillages dealt with straight away? ☐
- Is equipment stored to minimise the risk of it falling? ☐
- Are windows and glazing marked and made of safety material when necessary for reasons of health and safety? ☐
- Can windows, skylights, etc. be operated safely? ☐
- Can windows, etc. be cleaned safely? ☐
- Are traffic and emergency routes kept clear? ☐
- Do doors or gates present any hazard? ☐
- Is drinking-water labelled? ☐

Photocopiable resource

CHECKLIST No. 13 cont.

- Is storage for clothes, bags, etc. acceptable? ☐
- Are trailing leads kept to a minimum and if used, kept in rubber strips or taped down? ☐

COPING AT THE TIME OF AN INCIDENT

- First aid. ☐
- Isolate any hazardous equipment or building feature, etc. ☐
- Keep others safe. ☐

FOLLOW-UP

- Investigate, report, record, review risk assessment ☐

EXTRA POINTS FOR YOUR SCHOOL

NAME AND TELEPHONE NO. OF EXTERNAL CONTACT FOR HELP AND/OR ADVICE

Photocopiable resource

CHECKLIST No. 14

LETTINGS (ALSO RELEVANT TO PTA FUNCTIONS)

ACTIONS TO MAKE INCIDENTS LESS LIKELY AND/OR LESS SEVERE

It is normally agreed that you need to make sure that your premises are suitable for the letting in question. The person leasing the activities is normally held to be responsible for what actually happens during the session. It is therefore useful for a school to *make a number of points clear* to people hiring premises. Many caretakers have found it helpful to give such people a checklist for them to sign and to be kept on file in school. You can easily make such a checklist from the points given below.

Whilst the points given are relevant to PTA functions, it could be held that PTAs, or similar, are part of the school and so would not be seen as a separate entity should an investigation occur. Heads and governing bodies are urged to be satisfied with the health and safety implications of any activity organised by the PTA before it goes ahead.

- Is it clear what actual space has been made available for use, e.g. is it just the hall or the hall plus the staff room? ☐
- Does the person letting the premises know:
  - where the fire extinguishers are? ☐
  - where the fire alarm call points are? ☐
  - where the first aid box is located? ☐
  - where the emergency exits are sited? ☐
  - where the toilets are? ☐
  - how to get access to a telephone? ☐
- Is it made clear what the smoking policy is? ☐
- Has it been made clear to the person in charge that he or she must arrange emergency evacuation procedures? (They might need a torch for emergencies.) ☐

Photocopiable resource

CHECKLIST No. 14 cont.

- Does the person in charge know that he or she must report any incidents or damage to the caretaker before they leave the site? ☐
- Does the person in charge understand that he or she must explain all of these arrangements to the members of their group? ☐

COPING AT THE TIME OF AN INCIDENT

- This will be the responsibility of the person in charge of the group working in collaboration with your caretaker if he or she is on site. The ability to conduct a situational risk assessment would help here. ☐

FOLLOW-UP

- It is necessary to follow up carefully to identify if the school could reasonably have done more to prevent an incident or to provide emergency facilities. ☐
- If the incident was exclusively a result of the arrangements made by the leader (e.g. poor karate instructions) then that person will need to take action. ☐
- If the incident placed the school, staff or children at a level of risk you are not willing to accept, then unless adequate assurances are obtained, consider terminating the letting. ☐
- Investigate, report, record, review risk assessment. ☐

EXTRA POINTS FOR YOUR SCHOOL

NAME AND TELEPHONE NO. OF EXTERNAL CONTACT FOR HELP AND/OR ADVICE

Photocopiable resource

CHECKLIST No. 15

MAINTENANCE (PREMISES)

ACTIONS TO MAKE INCIDENTS LESS LIKELY AND/OR LESS SEVERE

- How do you carry out and record regular premises checks? ☐
- How do staff report any premises concerns? ☐
- How is this recorded and acted upon? ☐
- Who is responsible for chasing up work? ☐
- Who is responsible for agreeing that work done is satisfactory? ☐
- What jobs are the caretakers competent to do themselves? ☐
- Is there a list of these jobs? ☐

COPING AT THE TIME OF AN INCIDENT

- When reporting faults, do staff also make the area 'safe'? This will vary with circumstances and could include signs, cordoning off areas, special supervision, taking out of use, timetable changes, etc. ☐
- Should a fault suddenly occur, e.g. falling guttering, then the points made for 'reporting faults' (above) apply here. ☐
- First aid might be necessary. ☐

Photocopiable resource

CHECKLIST No. 15 cont.

FOLLOW-UP

- Check for associated problems. ☐
- Advise your LEA so that they can advise other schools. ☐
- Discuss whether action to make 'safe' was too much or too little. ☐
- Investigate, report, record, review risk assessment. ☐

EXTRA POINTS FOR YOUR SCHOOL

NAME AND TELEPHONE NO. OF EXTERNAL CONTACT FOR HELP AND/OR ADVICE

Photocopiable resource

CHECKLIST No. 16

MAINTENANCE (EQUIPMENT)

ACTIONS TO MAKE INCIDENTS LESS LIKELY AND/OR LESS SEVERE

- Do you keep a list of equipment that requires regular maintenance or testing, and does it record dates of maintenance or testing and who carried it out? (Include portable electrical appliances, gymnastic equipment, fire extinguishers, alarms, etc.) ☐
- Do staff regularly check the day-to-day equipment and furniture in their rooms? ☐
- Are staff and children trained to use equipment? ☐
- Do staff check certain equipment before use, e.g.
  - audio-visual aids ☐
  - PE equipment? ☐
- How do staff report any concerns relating to equipment? ☐
- How is this recorded and acted upon? ☐
- Who is responsible for chasing up work? ☐
- Who is responsible for agreeing that it is satisfactory? ☐
- What jobs are the caretakers competent to do themselves? ☐
- Is there a list of these jobs? ☐
- Is it clear that staff may bring their own equipment on site only with permission? ☐

Photocopiable resource

CHECKLIST No. 16 cont.

COPING AT THE TIME OF AN INCIDENT

- When reporting faults, do staff also make the item 'safe'? This will vary with circumstances and could include signs, cordoning off large items, special supervision, taking out of use, etc. ☐
- Should a fault suddenly occur, then the points made for 'reporting faults' (above) apply here. ☐
- First aid might be necessary. ☐

FOLLOW-UP

- Check for associated problems. ☐
- Advise your LEA so that they can advise other schools. ☐
- Discuss whether action to make 'safe' was too much or too little. ☐
- Investigate, report, record, review risk assessment. ☐

EXTRA POINTS FOR YOUR SCHOOL

NAME AND TELEPHONE NO. OF EXTERNAL CONTACT FOR HELP AND/OR ADVICE

Photocopiable resource

CHECKLIST No. 17

OPEN EVENINGS

ACTIONS TO MAKE INCIDENTS LESS LIKELY AND/OR LESS SEVERE

- Have you considered how much access is necessary, to what areas and what facilities? This needs to be balanced against security, personal safety and emergency evacuation needs. ☐
- Are any staff at risk due to the isolation of their location? Could they and their children's work, etc. be located in a safer place? ☐
- Are any staff at risk due to any other hazards? ☐
- Are there clear evacuation procedures on which the staff have been briefed? ☐

COPING AT THE TIME OF AN INCIDENT

- In the event of an incident occurring which requires emergency evacuation, staff will need to escort parents to the assembly points, checking specified locations and clearing public areas as they go. It is extremely unlikely that you will know who is on the site at any one time, so the emphasis will be on trying to ensure complete evacuation. ☐
- Staff will also need to be aware of potential personal safety situations and to be aware of the signs of such incidents developing so that they can, for instance, try to calm the situation, make an excuse and leave the room or raise the alarm. ☐
- First aid. ☐

Photocopiable resource

CHECKLIST No. 17 cont.

FOLLOW-UP

- ■ Investigate, report, record, review risk assessment. ☐

EXTRA POINTS FOR YOUR SCHOOL

NAME AND TELEPHONE NO. OF EXTERNAL CONTACT FOR HELP AND/OR ADVICE

Photocopiable resource

CHECKLIST No. 18

PLAYGROUND EQUIPMENT

ACTIONS TO MAKE INCIDENTS LESS LIKELY AND/OR LESS SEVERE

- Is it designed for its purpose, including the amount of regular heavy use that it will receive? ☐
- Can children get off safely rather then have to jump off? ☐
- Is it located far enough away from other equipment, the school building, etc.? ☐
- Is it located so that it can easily be seen from the school building (to discourage unauthorised use)? ☐
- Is a suitable safe surface provided under and around the equipment? ☐
- Are there signs to indicate any restrictions on use? ☐
- Do you have rules for use? For instance:
  - numbers ☐
  - behaviour ☐
  - supervision ☐
  - medical considerations ☐
  - special needs considerations ☐
  - weather conditions ☐
  - maximum age range at any one time ☐
  - clothing/footwear ☐
  - not for use before or after school. ☐

Photocopiable resource

CHECKLIST No. 18 cont.

- Is the equipment checked daily? For instance, for: ☐
  - litter, glass, dog faeces ☐
  - damage to any safe surface ☐
  - damage to the equipment ☐
  - rigidity ☐
  - pieces missing (including tube end covers) ☐
  - horizontal bars loose ☐
  - cracks, splinters, splits (depending on material) ☐
  - bolts, screws loose or poking out ☐
  - sharp edges through use or damage ☐
  - vandalism – insertion/addition of objects ☐
  - alignment of components ☐
  - handrails secure ☐
  - steps secure ☐
  - chains or ropes not damaged or twisted ☐
  - no excessive wear, deterioration, etc. ☐

  These are general points; some caretakers have used them as a basis for itemised checklists for their particular pieces of equipment.

COPING AT THE TIME OF AN INCIDENT

- First aid. ☐
- Supervision of other pupils. ☐
- Making equipment/location 'safe'. ☐

Photocopiable resource

CHECKLIST No. 18 cont.

FOLLOW-UP

- If an accident is due to a fault/failure of the equipment, it is useful to take photographs for possible future reference. ☐
- Do all you can to take the equipment out of use and make safe. ☐
- If the manufacturer offers immediate replacement of faulty parts, take advice before agreeing how this is to be done. ☐
- Investigate, report, record, review risk assessment. ☐

EXTRA POINTS FOR YOUR SCHOOL

NAME AND TELEPHONE NO. OF EXTERNAL CONTACT FOR HELP AND/OR ADVICE

Photocopiable resource

CHECKLIST No. 19

PLAYGROUND SAFETY

See also Chapter 3, 'Playtimes'.

ACTIONS TO MAKE INCIDENTS LESS LIKELY AND/OR LESS SEVERE

■ Procedures

- Do you have written supervision procedures for staff to follow? ☐

■ Playground

- surface state? ☐
- cleaning? ☐
- plants and bushes (physical hazards and poisoning)? ☐
- damaged fences? ☐
- hazardously located windows and other building features? ☐

■ Equipment

See Checklist 18.

■ Supervision

- ratio? ☐
- location? ☐
- fixed site or patrolling? ☐
- are the three above related to risk assessment? ☐
- any 'blind spots'? ☐
- clarity about rules for behaviour? ☐
- acceptable games? ☐
- when can children be 'sent in'? ☐
- when should help be summoned? ☐
- supervisors aware of danger signs in behaviour? ☐

Photocopiable resource

CHECKLIST No. 19 cont.

– rules about toys, ball sizes, etc? ☐
– procedures regarding visitors? ☐
– attitude to 'visiting parents'? ☐

■ Before and after school

– ten minutes of supervision provided? ☐
– first aid cover provided? ☐
– ball games? ☐
– use of equipment? ☐
– parents informed? ☐

COPING AT THE TIME OF AN INCIDENT

■ First aid. ☐

■ Safety of others. ☐

■ Isolation of hazardous areas. ☐

FOLLOW-UP

■ Short-term changes to arrangements, use of space, supervision, etc. ☐

■ Investigate, report, record, review risk assessment. ☐

EXTRA POINTS FOR YOUR SCHOOL

NAME AND TELEPHONE NO. OF EXTERNAL CONTACT FOR HELP AND/OR ADVICE

Photocopiable resource

CHECKLIST No. 20

PONDS

This list applies to small ponds (shallower than 0.5 metre) where it is possible to wade in to rescue someone.

ACTIONS TO MAKE INCIDENTS LESS LIKELY AND/OR LESS SEVERE

- Is the pond designed so that edges can be seen clearly? ☐
- Are the edges safe to stand on? ☐
- Does vegetation obscure the edge or give the impression of a firm surface? ☐
- Is there enough space around the pond for it to be used safely? There needs to be enough space for children to work without the risk of being knocked in. ☐
- Is the pond area fenced off in some way? If you are growing a hedge, you will need a temporary fence. Avoid fences that act as climbing frames. ☐
- Is there a sign warning visitors of the pond area? ☐
- Are the pond and pond area regularly checked for tin cans and other litter? ☐
- Do users wear suitable footwear? ☐

Photocopiable resource

CHECKLIST No. 20 cont.

- Hygiene:
  - Keep tools and equipment in good condition (no rust). ☐
  - Cover cuts and abrasions with waterproof dressings. ☐
  - Use plastic gloves or bags to cover hands when collecting specimens. ☐
  - Collect insects, etc. with brushes or pooters. ☐
  - Keep nests, unwashed feathers in sealed polythene bags. ☐
  - Everyone working in and around the pond or with samples must wash hands and forearms afterwards. ☐
- Ensure that you are satisfied with supervision arrangements. ☐

COPING AT THE TIME OF AN INCIDENT

- First aid. ☐
- Safety of others. ☐
- It is useful to have some dry clothes on the premises. ☐
- Isolation of pond area. ☐

Photocopiable resource

CHECKLIST No. 20 cont.

FOLLOW-UP

- Always tell parents if a child has fallen in and/or, swallowed pond water and/or sustained a cut or abrasion when working in and around the pond and associated activities. Parents need to be advised that they might wish to consult a doctor (tetanus, Weil's disease, etc.). ☐
- Continue non-use of pond area until you are satisfied that it is acceptable to resume. ☐
- Investigate, report, record, review risk assessment. ☐

EXTRA POINTS FOR YOUR SCHOOL

NAME AND TELEPHONE NO. OF EXTERNAL CONTACT FOR HELP AND/OR ADVICE

Photocopiable resource

CHECKLIST No. 21

TRAINING

ACTIONS TO MAKE INCIDENTS LESS LIKELY AND/OR LESS SEVERE

■ Have you considered the necessary content of training? ☐

- needs of school and risk assessment ☐
- LEA and/or legal requirements ☐

■ Have you considered who needs training? ☐

- whole staff needs ☐
- group needs ☐
- individual needs ☐

■ How do you provide training? ☐

- about health and safety ☐
- about particular topics ☐
- to impart particular skills ☐
- to support curriculum delivery ☐
- to meet specialist needs ☐

■ How do you record training?

- individual records ☐
- topic records ☐
- keeping track of renewal dates ☐

Photocopiable resource

CHECKLIST No. 21 cont.

COPING AT THE TIME OF AN INCIDENT

Not relevant – if something goes wrong due to lack of training, then apart from normal emergency procedures, first aid, reporting, etc., little else will help at the time.

FOLLOW-UP

- At least once a year, review your training arrangements. ☐
- Assess training needs of all new staff. ☐
- Discuss training needs as part of regular consultations. ☐
- Reinforce these processes by the use of the safety records that you have been keeping. ☐

EXTRA POINTS FOR YOUR SCHOOL

NAME AND TELEPHONE NO. OF EXTERNAL CONTACT FOR HELP AND/OR ADVICE

Photocopiable resource

CHECKLIST No. 22

VISITS NEAR SCHOOL

This checklist is not meant to address day visits to remote places of interest and certainly not overnight stays or adventure activities.
It is, within the spirit of the book, meant to address 'day-to-day' visits to parks, museums, libraries, etc.

ACTIONS TO MAKE INCIDENTS LESS LIKELY AND/OR LESS SEVERE

Some points to consider:

- purpose of visit ☐
- nature of location(s) ☐
- nature of planned activities ☐
- route to and from the location ☐
- means of transport ☐
- boarding and alighting points for buses ☐
- ages of children ☐
- number of children ☐
- competence of children ☐
- any special or medical needs ☐
- weather conditions ☐
- time of day. ☐

Photocopiable resource

CHECKLIST No. 22 cont.

Thinking about the points listed above:

- How much back-up do you have at school to cope with emergencies? If there are two trips out, and both have problems, can you provide help without creating a potentially hazardous situation in school? ☐

- How many supervisory staff are needed? ☐

  For local visits, walks, etc., in normal circumstances the DfEE advises:

  1 adult per 6 pupils for years 1 to 3 (higher ratio for younger pupils)

  1 adult per 10 to 15 pupils for years 4 to 6

- What proportion of supervisors need to be teachers? ☐

- What proportion need to be members of school staff? ☐

- Are all staff briefed about what to do under normal circumstances and in an emergency? ☐

- What skills do supervisory staff need (e.g. first aid)? ☐

- Do you have first aid bags and provision for other emergencies (sick bucket or bags, changes of clothes, water and cups, yellow bags for clinical waste, cloths for surfaces, wipes for skin)? ☐

- Will you provide written notes for helpers? ☐

Photocopiable resource

CHECKLIST No. 22 cont.

COPING AT THE TIME OF AN INCIDENT

- Who will take charge? ☐
- Who will take charge if that person is injured? ☐
- Can both people be expected to assess the situation and decide what to do? (Conduct a situational risk assessment.) ☐
- Can any injured person be properly cared for while the rest of the group is adequately supervised? ☐
- How will the emergency services and/or the school be informed of what is happening? ☐

FOLLOW-UP

- Investigate, report, record, review risk assessment. ☐

EXTRA POINTS FOR YOUR SCHOOL

NAME AND TELEPHONE NO. OF EXTERNAL CONTACT FOR HELP AND/OR ADVICE

Photocopiable resource

CHECKLIST No. 23

VEHICLES ON SITE

ACTIONS TO MAKE INCIDENTS LESS LIKELY AND/OR LESS SEVERE

It is always best to separate pedestrians from vehicles. If this is not possible, you will need to think about separating the times in which vehicles and pedestrians use certain routes. If this is not practicable, e.g. delivery of school meals during times when your playground is in use, then extra vigilance is required by those supervising children.

- Do vehicles that are authorised to park on your site have window stickers indicating that they have such authorisation? ☐
- Is car parking available for authorised visitors? Is this clearly indicated? ☐
- If delivery drivers can make deliveries only at certain times of the day, is this made clear on order forms? ☐
- Do you remind parents every term about where they can and cannot park on your school site? ☐
- Do you have signs indicating a speed limit on your site? ☐
- Have you identified access routes for emergency vehicles and are these kept clear at all times? ☐
- Have you identified emergency evacuation routes, especially spaces outside fire exit doors, and are these kept clear at all times? ☐
- Is there a safe place for coaches and buses to load and unload? ☐

Photocopiable resource

CHECKLIST No. 23 cont.

COPING AT THE TIME OF AN INCIDENT

- If there is an accident, then first aid and/or the help of emergency services could be necessary. ☐
- Locations need to be taken out of use. ☐
- Pedestrian access will need temporary modification to keep people away from the incident location. ☐

FOLLOW-UP

- Investigate, report, record, review risk assessment. ☐
- As this area involves an interface with many users, e.g. parents and contractors, consider that publicity needs to be given to revised procedures. ☐

EXTRA POINTS FOR YOUR SCHOOL

NAME AND TELEPHONE NO. OF EXTERNAL CONTACT FOR HELP AND/OR ADVICE

Photocopiable resource

CHECKLIST No. 24

WASTE DISPOSAL

ACTIONS TO MAKE INCIDENTS LESS LIKELY AND/OR LESS SEVERE

- Hazardous substances
  - Does everyone know that they must dispose of these substances in accordance with the information provided by the manufacturer and/or your arrangements to comply with the Control of Substances Hazardous to Health Regulations? (See Checklist 12.) ☐
- Items with sharp edges (broken glass or china, nails, blades, etc.):
  - Do the children know that these must be put into a bin only with the knowledge of an adult? ☐
  - Before putting such items in bins, they need to be wrapped up in, say, newspaper and labelled (there have been accidents where cleaners and caretakers have cut their hands when emptying bins containing such items). ☐
- Waste paper – to reduce the risk of arson:
  - Are all waste-paper bins emptied at the end of the school day (i.e. not left overnight thus providing a potential bonfire for an intruder)? ☐
  - Is all waste paper placed in a suitable lidded bin, if possible 10 metres away from the school building and chained to the ground? ☐
- Body fluids:
  - You will need to follow current advice provided by either your LEA or your local consultant in communicable disease control (usually contactable through the Health Authority). ☐

Photocopiable resource

CHECKLIST No. 24 cont.

- Left-over items from school and PTA sales
  - These will need to be stored safely or removed from the site as quickly as possible. ☐

COPING AT THE TIME OF AN INCIDENT

Depending on the incident, first aid may be necessary.

- If hazardous substances:
  - If substances are left out, do not leave them or the children unattended. Send for another adult to take them away to a safe location. ☐
  - If there is a spillage, keep people well away and follow the manufacturer's advice (taken from the information/product data sheets). This might include calling the emergency services. ☐
  - If anyone is hurt, follow the manufacturer's advice (taken from the information/product data sheets) and seek help as appropriate. This might include calling the emergency services. ☐
- If fire:
  - Follow fire procedures. ☐
- If body fluids:
  - Follow advice of consultant in communicable disease control. ☐

Photocopiable resource

CHECKLIST No. 24 cont.

FOLLOW-UP

- Investigate, report, record, review risk assessment. ☐

EXTRA POINTS FOR YOUR SCHOOL

NAME AND TELEPHONE NO. OF EXTERNAL CONTACT FOR HELP AND/OR ADVICE

Photocopiable resource

CHECKLIST No. 25

WORK EXPERIENCE PLACEMENTS

This section gives you a list of points to explain to work experience placement students from other schools. You will also find it useful for students from teacher training courses, though of course, you will want to modify the content, taking into account the maturity and experience of the older person.

ACTIONS TO MAKE INCIDENTS LESS LIKELY AND/OR LESS SEVERE

- Does the student know who they report to and who they should report to if this person is unavailable for any reason? ☐
- Are students clear that they must let someone know when they arrive and when they leave the site (including lunch breaks)? ☐
- Has the student been taken through the relevant points or your school's health and safety documentation? ☐
- Do they know if there are any areas in your school which you regard as being prohibited to them (e.g. the boiler room or the school meals kitchen)? ☐
- Are they clear about which work equipment they can and cannot use? ☐
- Have you shown them how to use the equipment that they are to use, and are you confident that they can use it safely (this includes, of course, PE equipment, games equipment, electrical equipment, etc.)? ☐

Photocopiable resource

CHECKLIST No. 25 cont.

- Have you pointed out to them that the correct equipment must always be used for the task in hand (e.g. they must not stand on desks or chairs, but always use the steps that you have provided for them)? ☐

- Hazardous substances:

  – If the students are going to use any hazardous substance, have you explained to them the precautions that they must take? ☐

- Lifting and carrying:

  – Have you pointed out to them the correct way to move (or not move) heavy or awkwardly shaped items? ☐

- Emergency procedures – Do they know:

  – where the first aid equipment is kept? ☐
  – who the first-aider is? ☐
  – how to report accidents? ☐
  – how to sound the fire alarm? ☐
  – what to do if the fire alarm sounds? ☐
  – what they are to do in any other emergency situation? ☐

- Medical emergencies:

  – do they know what you want them to do in such situations? ☐

COPING AT THE TIME OF AN INCIDENT

- Depending on the nature of the incident, required emergency procedures need to be followed. ☐

Photocopiable resource

CHECKLIST No. 25 cont.

FOLLOW-UP

- Investigate, report, record, review risk assessment. ☐

EXTRA POINTS FOR YOUR SCHOOL

NAME AND TELEPHONE NO. OF EXTERNAL CONTACT FOR HELP AND/OR ADVICE

Photocopiable resource

CHECKLIST No. 26

WORKING ALONE

Increasingly it seems to be the case that members of teaching and non-teaching staff work alone. This is especially true of the smallest schools.

ACTIONS TO MAKE INCIDENTS LESS LIKELY AND/OR LESS SEVERE

- Can work be scheduled to reduce lone working to a minimum? ☐
- Does someone know that the person is working alone? ☐
- Is there a clear understanding about what can or cannot be done by lone workers? e.g.: ☐
  - using ladders, steps, etc. ☐
  - using potentially hazardous equipment ☐
  - interviewing potentially difficult people. ☐
- Can the lone worker summon help if needed? ☐
- Is first aid equipment at hand (not locked away)? ☐
- Is the lone worker clear that he or she must secure the building on leaving? ☐

Photocopiable resource

CHECKLIST No. 26 cont.

COPING AT THE TIME OF AN INCIDENT

- Lone workers need to be competent to cope with a whole range of possibilities. ☐
- In some circumstances, a lone worker will need to be able to conduct a 'situational risk assessment' to decide what to do in an incident, e.g. approach an intruder or summon assistance. ☐

FOLLOW-UP

- Investigate, report, record, review risk assessment. ☐

EXTRA POINTS FOR YOUR SCHOOL

NAME AND TELEPHONE NO. OF EXTERNAL CONTACT FOR HELP AND/OR ADVICE

Photocopiable resource

BLANK TEMPLATE

TOPIC . . . .

ACTIONS TO MAKE INCIDENTS LESS LIKELY AND/OR LESS SEVERE

COPING AT THE TIME OF AN INCIDENT

FOLLOW-UP

EXTRA POINTS FOR YOUR SCHOOL

NAME AND TELEPHONE NO. OF EXTERNAL CONTACT FOR HELP AND/OR ADVICE

Chapter 6 Some wider issues

This section outlines some considerations which, whilst not strictly within the scope of the book, are closely related in terms of school management. These 'wider issues' are:

1. Advice
2. Consultation
3. Curriculum safety
4. Medical facilities and welfare
5. Personal safety and security
6. Pupils (and staff) with special needs
7. School transport
8. Stress

Advice

Employers have to appoint one or more competent persons to assist them in complying with statutory requirements. Such competent persons might be employees or (in the case of a small firm) the employer themselves. Where there is not sufficient competence, then the employer needs to bring in an external provider of some sort to help. In very many cases, businesses operate by using a mixture of both.

Very often in schools, advice will come from senior staff (including the headteacher). The Guidance to the 1999 Management Regulations indicates that for 'simple situations', someone needs to understand current relevant best practice, be aware of the limits of their own knowledge and experience and be willing to supplement this with expert help and advice when necessary. For 'more complicated' situations, the Guidance recommends obtaining the services of someone with a higher level of knowledge and experience. So within a school, day-to-day advice could be the responsibility of the relevant post holder. LEA maintained schools would probably then refer to LEA advisory services (not only Health and Safety but also Curriculum, Premises, Health, etc.) for advice with the more complicated situations. Other schools will need to decide where to source advice.

However you decide to do this, it is best to mention it in two places in your documentation:

1. in the Organisation section when you are saying who is responsible for what, and
2. in the Arrangements section when you describe how you provide advice.

(You could save time by cross-referencing.)

Consultation

It is essential to consult with all staff over matters of health and safety. It is also very useful to discuss issues with the children. This raises their awareness and understanding of a literally vital area of life and also brings fresh insights into your own management. As health and safety is of necessity part of all aspects of school life, you will need to include it as a consideration in the discussion of all activities – and keep a note that you have done so.

In addition to making health and safety issues part of general decision making, you will need to make arrangements for formal consultation. If trade unions in your school have appointed safety representatives to represent their members' interests, there needs to be regular formal consultation with them over a range of issues. If there is a properly established safety committee, then this will need to meet regularly. Meetings with safety representatives, and of safety committees, need proper minutes. If you do not have formal arrangements, it is a good idea to include health and safety as a regular item in meetings with all staff (not just teachers) and keep minutes of what was discussed. The booklet *Safety representatives and safety committees* is obtainable from HSE Books; the address is in Chapter 7.

As with advice, consultation would benefit from being recorded in both the Organisation and the Arrangements sections of your documentation.

Curriculum safety

Teaching notes need to include references to any significant hazards that could harm the children or the teacher, together with how the risk of harm occurring is to be minimised. In addition, notes need to include material on how children are to be taught to carry out activities in safe ways (OFSTED looks out for this). *Safety across the curriculum* is a useful resource book (see Chapter 8). It is also worth looking at publications of the Association for Science Education, British Association of Advisers and lecturers in Physical Education and the Consortium of Local Education Authorities for the Provision of School Science (see Chapter 7).

Medical facilities and welfare

The DfEE publication *Supporting Pupils with Medical Needs* and Circular 14/96 are good references. It is also worth contacting the Health Education Authority which publishes all sorts of useful materials (e.g. *The Use of Sunscreens in School: a good practice guide*). Most medical conditions have an associated support group. Advice about communicable diseases is usually obtainable via your local Health Authority's Consultant in Communicable Disease Control.

Please remember that medical and welfare facilities are needed for staff as well as pupils.

Personal safety and security

In his report on the Public Enquiry into the events at Dunblane Primary School, Lord Cullen made it clear that getting the physical security of a building right did not necessarily make the people safe and vice versa. Nevertheless, both physical security and personal safety need to be the subject of risk assessment. There are a number of helpful books and resources listed in Chapter 8. A simple risk assessment sequence would go something like this:

1. Identification of areas for consideration
   Consider work activities, specific groups of people, specific locations on site, building features, times of day, week, year, storage of records, pupil/teacher work, valuable equipment, etc.
2. Consideration of hazards
   These could be fire, assaults, water damage, thefts, vandalism, etc.
3. Think about how each hazard could affect your areas for consideration. For instance, a fire in a store room could destroy work required for assessment.
4. Describe the arrangements that are currently in place to safeguard against the effect(s) of the hazard(s).
   It is helpful to think in terms of three areas of provision:
   a Technology, e.g. building design, locks, alarms, fencing
   b Management systems, e.g. signing-in systems, arrangements for lone workers, parental interviews of all sorts, handling cash
   c Human factors, e.g. the commitment of staff, pupils, governors; training; individual competence; how well people follow systems; how well they use the technology.
5. Form a judgement, on the basis of experience in your school and any other relevant situation, about the adequacy of your arrangements.
   If they appear adequate, continue making sure that they operate correctly and keep records of any incidents; if not, go to item 6.
6. For each area that you feel is inadequate, decide whether residual risk is high, medium or low.
   Sort the inadequacies into order, highest risk at the top of the list, lowest at the bottom. This is a risk assessed priority schedule.
7. Prepare an action plan on the basis of the schedule.
8. Implement your action plan.

Pupils (and staff) with special needs

Taking into account the effect of competency on risk assessment, it is clear that pupils who have special needs will need to be considered when you carry out risk assessment. This applies also to pupils with medical needs, as well as to staff with both special needs and medical needs. In some cases e.g. pupils who have statements of special education need, it is possible to add safety requirements to the statement, e.g. the need for assistance with certain aspects of moving around. In this way the safety consideration stays with the pupil's records and hopefully would be taken into account when assessing necessary resources.

School transport

Only a few nursery and primary schools have their own transport or hire minibuses. RoSPA and the Community Transport Association publish *Minibus Safety: A Code of Practice* – details in Chapter 8.

Stress

As with personal safety and security, there are references to publications about stress in Chapter 8. Once again, a risk assessed approach is needed. A simple sequence would be:

1. Either as part of general risk assessment, or as a specific activity, identify those hazards which are causing stress within your school. Hazards, for instance, could be organisational, interpersonal, environmental.
2. Either as part of item 1, or in association with it, create opportunities for staff (and pupils if necessary) to talk about how they feel and how things could be better. This needs to be done in an open, supportive way.
3. Think about ways of removing or reducing hazards that lead to stress. Some of these will be related to how the school is organised, others to how people work together in groups and still others to how individuals approach the challenges of their work or provided activities.
4. Provide training and, if necessary, assistance to help individuals either reduce the effect of stressful situations on them or cope with the effects of stress.

Chapter 7 Helpful organisations

Association for Science Education, College Lane, Hatfield, Herts AL10 9AA; tel: 01707 267411

British Association of Advisers and Lecturers in Physical Education (BAALPE), Nelson House, 6 The Beacon, Exmouth, Devon EX8 2AG; tel: 01395 263247

Child Accident Prevention Trust, 4th Floor, Clerks Court, 18–20 Farringdon Lane, London EC12 3AU; tel: 020 7608 3828

Consortium of Local Education Authorities for the Provision of School Science (CLEAPSS), Brunel University, Uxbridge UB8 3PH; tel: 01895 251496

DfEE Publications Centre, PO Box 5050, Sudbury, Suffolk CO10 6ZQ; tel: 0845 6022260; fax: 0845 6033360

Health Education Authority (HEA), Health Promotion Information Centre, Hamilton House, Mabledon Place, London WC1H 9TX; tel: 020 7383 3833

Health and Safety Executive Information Centre, PO Box 1999, Broad Lane, Sheffield S3 7HQ; infoline: 0541 545500

HSE Books, PO Box 1999, Customer Services Dept, Sudbury, Suffolk CO10 6FS; tel: 01787 881165; fax: 01787 313995

National Playing Fields Association (NPFA), 25 Ovington Square, London SW3 1LQ; tel: 020 7584 6445

Royal Society for the Prevention of Accidents (RoSPA), Egbaston Park, 353 Bristol Road, Birmingham B5 7ST; tel: 0121 972 2000

The Stationery Office, The Publications Centre, PO Box 276, London SW8 5DT; tel: 020 7873 9090 (orders); tel: 020 7873 0011 (enquiries)

Chapter 8 Useful books and resources

Child Accident Prevention Trust, 4th Floor, Clerks Court, 18–20 Farringdon Lane, London EC12 3AU; tel: 020 7608 3828

Accident prevention in day care and play settings: a practical guide – This book is exactly what it says it is and contains lots of useful, relevant and practical advice for safety when working with younger children.

Getting over an accident (series) – Five leaflets: Advice for children aged 8 and under; Advice for children over 8; Advice for young people; Advice for parents and carers; Guidelines for professionals. This material was produced as a practical response to *Healing the hidden hurt* (see below).

Healing the hidden hurt (the emotional effects of children's accidents) – The findings of a study investigating the need for and the provision of emotional support to children who were treated for accidental injuries at Accident and Emergency departments.

Keeping Kids Safe – activities and information for parents' groups – A booklet full of ideas and information for all who care for young children (also has ideas which can be adapted for use in class).

The Risk Pack – a teachers' resource to help children assess and manage risk – Based on the principle that if children understand risk and risk assessment, they are more likely to keep safe. The book is aimed primarily at Key Stage 2 (though it can be adapted in use) and contains: information on the process of assessing risk; a flow chart identifying the steps; objectives which can be met by the use of the materials; ideas for topics, themes and activities for pupils.

Department for Education and Employment, DfEE Publications Centre, PO Box 5050, Sudbury, Suffolk CO10 6ZQ; tel: 0845 6022260; fax: 0845 6033360

Guidance on First Aid for Schools – A very straightforward and easy-to-use description of what schools need to do to respond to the requirement to provide first aid. Written in consultation with teacher unions, LEAs, health services and voluntary organisations.

Health and Safety of Pupils on Educational Visits – The second paragraph contains the sentences 'The booklet does not seek to replace local or other professional guidance or regulations. Where appropriate, LEAs should be the first source of advice.' Educational visits are a massive area for consideration – this booklet reinforces my view. It does give an overview of what is involved, but you will also need to obtain materials from other sources together with advice from your LEA (if you are working in an LEA maintained school).

Improving Security in Schools – A very practical little booklet giving clear and helpful advice.

Personal Safety and Violence in Schools – A report prepared by Leicester University, commissioned by the Suzy Lamplugh Trust, largely funded by the DfEE. The report contains lots of interesting information and in particular a section titled 'Towards Good Practice' which indicates what schools and others could do to help.

School Security, Dealing with Troublemakers – Published by the DfEE and the Home Office, offers guidance to the law relating to troublemakers in and around schools.

Supporting Pupils with Medical Needs – Guidance to help schools draw up policies on managing medication in schools, and to put in place effective management systems to support individual pupils with special needs. (Needs to be read with Circular 14/96 – Supporting Pupils with Medical Needs in School). A very helpful publication giving practical advice, specimen forms, etc.

Work Experience – a guide for employers – Primarily intended to assist employers in their provision of work experience placements, contains a useful section about health and safety.

HSC/HSE, HSE Books, PO Box 1999, Customer Services Dept, Sudbury, Suffolk CO10 6FS; tel: 01787 881165; fax: 01787 313995

Regulations are part of the law, you must follow them.

Approved Codes of Practice are approved by the Health and Safety Commission and give practical advice on following the law. You may use other methods to follow the law, but if you should be prosecuted and did not follow the Approved Code, you would need to demonstrate that you had complied with the law in some other way.

Guidance – If you follow Health and Safety Commission Guidance, you will normally be doing enough to comply with the law.

NB: These are summary definitions, full definitions are provided in the front of the relevant documents.

Adventure activities: five steps to risk assessment – Very clearly written booklet which relates risk assessment to adventure activities and includes the helpful note that risk assessments must be 'suitable and sufficient – not perfect!' (The phrase also appears in *Five Steps to Risk Assessment* – see below.)

Display screen equipment work – Contains the regulations together with guidance for use. Whilst these relate primarily to requirements with regard to employees, they will also help you to decide how to reduce risks to children.

Five Steps to Risk Assessment – This new edition of a famous leaflet provides straightforward advice on risk assessment.

Health and safety guidance for school governors and members of school boards – This replaces *The responsibilities of school governors for health and safety*. It deals with the relationships between employers, governing bodies, employees and others. It does not deal with the responsibilities of 'persons who have to any extent control of premises' (which, it could be argued, governing bodies are).

Management of health and safety at work – Contains the Regulations and Approved Code of Practice about how health and safety must be managed. Updated in 1999, now includes material on young people and new and expectant mothers.

Managing health and safety in schools – Describes the Education Service Advisory Committee's view of the key elements of effective health and safety management in schools.

Managing occupational stress – A guide for managers and teachers in the school sector – Advice from the Education Services Advisory Committee about stress. NB: At the time of writing, the Health and Safety Commission was considering the preparation of a draft approved code of practice and guidance on work-related stress.

Personal protective equipment at work – Contains the Regulations and clear practical guidance on the provision and use of personal protective equipment. This sounds a bit 'industrial', but will apply to many adults in school (and some pupils).

Reporting of Injuries, Diseases and Dangerous Occurrences Regulations (A guide to the) – Contains Regulations and Guidance on how to follow the law. If you are an LEA school, you will find it more straightforward to go to LEA procedures. If not, you might feel more comfortable making a start with the leaflet *Everyone's guide to RIDDOR.*

Safe use of work equipment – Contains Regulations, Approved Code of Practice and Guidance on the Provision and Use of Work Equipment Regulations 1998. The point is made that 'The scope of "work equipment" is extremely wide' and gives among the examples hammers, photocopiers, and ladders. Many of the principles are relevant to schools, for instance regarding a climbing frame. Among the regulations are requirements for equipment to be suitable, properly maintained, inspected, risk assessed, and for those using and supervising equipment to be properly trained, informed and instructed.

Safety policies in the education sector – Advice from the Education Services Advisory Committee about the contents of school documentation.

Violence in the education sector – This replaces *Violence to staff in the education sector* and provides the Educations Service Advisory Committee's view on how the risk of violence can be managed in schools and other educational establishments.

Workplace health, safety and welfare – Regulations and Approved Code of Practice dealing with a whole range of premises issues including glazing, traffic routes, window cleaning and washing facilities – a much-referred-to document!

Merlin Communications (Education Desk), Dyer House, 3 Dyer Street, Cirencester, Gloucestershire GL7 2PP

Merlin publishes a range of risk assessment/health and safety distance learning and training materials including:

Personal Safety in Schools (produced with the Suzy Lamplugh Trust and members of the DfEE Working Group on School Security)

Property Risk Management for Schools (produced with Oxfordshire County Council)

The School Health and Safety Management Pack (produced with Derbyshire County Council)

Stress Management in Schools (produced with South Tyneside Council)

Other publications

Key Issues for Primary Schools (chapter on Health and Safety) – By Michael Farrell, published by RoutledgeFalmer 1999; Customer Orders Hotline: 08700 768853; www.routledgefalmer.com

Minibus Safety: A Code of Practice – RoSPA and the Community Transport Association. Details from: Royal Society for the Prevention of Accidents (RoSPA), Egbaston Park, 353 Bristol Road, Birmingham B5 7ST; tel: 0121 972 2000.

Safety across the curriculum – Edited by Carole Raymond, published by RoutledgeFalmer 1999; Customer Orders Hotline: 08700 768853; www.routledgefalmer.com

Should crisis call – Crisis Management in Schools: effective preparation a[n]d response – Written and published by Stirling Council, March 1999.

Wise before the event – coping with crises in schools – By William Yule an[d] Anne Gold, published by Calouste Gulbenkian Foundation, 1993.

Safety Across the Curriculum

Key Stages 1 and 2

Edited by **Carole Raymond**

'Throughout, the editor successfully treads the narrow path between being complacent and being alarmist . . . she offers primary teachers comprehensive guidance on safe practice across a range of curriculum areas . . . This book is a useful adjunct to safety advice from the DfEE.' – *Times Educational Supplement*

Addresses the needs of teachers faced with the increase in litigation and cases of educational malpractice. The book provides a framework for safe practice, examines the process of litigation and offers advice to make schools safer.

Contents: Part One: Creating a Framework for Safe Practice, *Carole Raymond.* Safety Politicies and Paperwork, *Carole Raymond.* Risk Assessment and the Management of Risk, *Peter Whittam.* **Part Two:** Safety in Design and Technology, *John Twyford* Health and Safety Regarding the Use of ICT in Schools, *Chris Taylor.* Physical Education, *Carole Raymond.* Learning in Safety and Learning About Safety: Issues in Practical Science, *Nigel Skinner.* Learning Lessons from Others' Experience: Safe Practice in the 'Outdoor Classroom', *Sue Thomas.* Organisations and Resources, *Carole Raymond.*

1999: 176pp
Pb: 0-7507-0984-7: £14.99